Proteomics: A Comprehensive Study of Proteins

Proteomics: A Comprehensive Study of Proteins

Edited by
Tanner Perry

Proteomics: A Comprehensive Study of Proteins
Edited by Tanner Perry
ISBN: 978-1-63549-239-2 (Hardback)

Published by Larsen and Keller Education,
5 Penn Plaza,
19th Floor,
New York, NY 10001, USA

Cataloging-in-Publication Data

Proteomics : a comprehensive study of proteins / edited by Tanner Perry.
p. cm.
Includes bibliographical references and index.
ISBN 978-1-63549-239-2
1. Proteomics. 2. Proteins. I. Perry, Tanner.
QD431 .P76 2017
572.6--dc23

Printed and bound in the United States of America.

For more information regarding Larsen and Keller Education and its products, please visit the publisher's website www.larsen-keller.com

Table of Contents

Preface

This book provides comprehensive insights into the field of proteomics. It talks extensively about the protein-protein interactions and their other applications and uses. Proteomics is concerned with the study of proteins. It refers to the study of proteome, which is a protein and examines its structure, composition, activities and intra-cellular protein interaction, etc. The book aims to shed light on some of the unexplored aspects of proteomic interaction. It is compiled in such a manner, that it will provide in-depth knowledge about the theory and practice of the subject. For all those who are interested in this topic, this text can prove to be an essential guide.

A foreword of all Chapters of the book is provided below:

Chapter 1 - The study of proteins is proteomics. Proteins are very important for living organisms. Entire sets of proteins are termed as proteomes. The chapter on protein interaction offers an insightful focus, keeping in mind the complex subject matter.; **Chapter 2** - Proteins are large molecules consisting of one or more chains of amino acid residues. Some of the aspects elucidated in this chapter are protein domain, protein biosynthesis, protein structure, protein structure prediction and enzymes. This is an introductory chapter that will introduce briefly all the significant aspects of proteins.; **Chapter 3** - The proteins studied in proteomics are cyanovirin-N, intrinsically disordered proteins, blood proteins and globular protein. Fundamentally disordered proteins are proteins that are not in a proper three-dimensional structure. This section has been carefully written to provide an easy understanding of the varied facets of proteins.; **Chapter 4** - The various interactive proteins domains discussed are SH2 domain, SH3 domain, Phosphotyrosine-binding domain, LIM domain, FERM domain, RNA-binding protein and pleckstrin homology domain. If the domain size of a protein is about 60 amino acids, it's known as SRC homology 3 domain. The diverse interactive protein domains have been carefully explained in this chapter.; **Chapter 5** - The method of separating biochemical mixtures is known as affinity chromatography whereas the chip-sequencing is the method used to analyze the interactions between proteins and DNA. Some of the other techniques discussed in the content are protein footprinting, microscale thermophoresis, two-hybrid screening, phage display and chromatin immuneprecipition.; **Chapter 6** - In molecular biology, the interactions that occur in a particular cell is known as an interactome whereas RNA interference is the process in which RNA molecules inhibit gene expression. This section elucidates the crucial aspects of the interactive network of proteins.; **Chapter 7** - Proteomics has a number of branches; some of the sub-disciplines are phosphoproteomics, degradomics and neuroprotemoics. Phosphoproteomics is the branch of proteomics that helps in the description and in the characterization of proteins. The chapter strategically encompasses and incorporates the major sub-disciplines of proteomics, providing a complete understanding.

I would like to thank the entire editorial team who made sincere efforts for this book and my family who supported me in my efforts of working on this book. I take this opportunity to thank all those who have been a guiding force throughout my life.

Editor

1

Understanding Proteomics and Protein Interaction

The study of proteins is proteomics. Proteins are very important for living organisms. Entire sets of proteins are termed as proteomes. The chapter on proteomics and protein interaction offers an insightful focus, keeping in mind the complex subject matter.

Proteomics

Proteomics is the large-scale study of proteins, particularly their structures and functions. Proteins are vital parts of living organisms, as they are the main components of the physiological metabolic pathways of cells. The term *proteomics* was first coined in 1997 to make an analogy with genomics, the study of the genome. The word *proteome* is a portmanteau of *prote*in and gen*ome*, and was coined by Marc Wilkins in 1994 while working on the concept as a PhD student.

Robotic preparation of MALDI mass spectrometry samples on a sample carrier.

The proteome is the entire set of proteins, produced or modified by an organism or system. This varies with time and distinct requirements, or stresses, that a cell or organism undergoes. Proteomics is an interdisciplinary domain that has benefited greatly from the genetic information of the Human Genome Project; it is also emerging scientific research and exploration of proteomes from the overall level of intracellular protein composition, structure, and its own unique activity patterns. It is an important component of functional genomics.

While *proteomics* generally refers to the large-scale experimental analysis of proteins, it is often specifically used for protein purification and mass spectrometry.

Complexity of the Problem

After genomics and transcriptomics, proteomics is the next step in the study of biological systems. It is more complicated than genomics because an organism's genome is more or less constant, whereas the proteome differs from cell to cell and from time to time. Distinct genes are expressed in different cell types, which means that even the basic set of proteins that are produced in a cell needs to be identified.

In the past this phenomenon was done by RNA analysis, but it was found not to correlate with protein content. It is now known that mRNA is not always translated into protein, and the amount of protein produced for a given amount of mRNA depends on the gene it is transcribed from and on the current physiological state of the cell. Proteomics confirms the presence of the protein and provides a direct measure of the quantity present.

Post-translational Modifications

Not only does the translation from mRNA cause differences, but many proteins are also subjected to a wide variety of chemical modifications after translation. Many of these post-translational modifications are critical to the protein's function.

Phosphorylation

One such modification is phosphorylation, which happens to many enzymes and structural proteins in the process of cell signaling. The addition of a phosphate to particular amino acids—most commonly serine and threonine mediated by serine/threonine kinases, or more rarely tyrosine mediated by tyrosine kinases—causes a protein to become a target for binding or interacting with a distinct set of other proteins that recognize the phosphorylated domain.

Because protein phosphorylation is one of the most-studied protein modifications, many "proteomic" efforts are geared to determining the set of phosphorylated proteins in a particular cell or tissue-type under particular circumstances. This alerts the scientist to the signaling pathways that may be active in that instance.

Ubiquitination

Ubiquitin is a small protein that can be affixed to certain protein substrates by enzymes called E3 ubiquitin ligases. Determining which proteins are poly-ubiquitinated helps understand how protein pathways are regulated. This is, therefore, an additional legitimate "proteomic" study. Similarly, once a researcher determines which substrates are ubiquitinated by each ligase, determining the set of ligases expressed in a particular cell type is helpful.

Additional Modifications

In addition to phosphorylation and ubiquitination, proteins can be subjected to (among others) methylation, acetylation, glycosylation, oxidation and nitrosylation. Some proteins undergo all these modifications, often in time-dependent combinations. This illustrates the potential complexity of studying protein structure and function.

Distinct Proteins are made under Distinct Settings

A cell may make different sets of proteins at different times or under different conditions, for example during development, cellular differentiation, cell cycle, or carcinogenesis. Further increasing proteome complexity, as mentioned, most proteins can undergo a wide range of post-translational modifications.

Therefore, a "proteomics" study can quickly become complex, even if the topic of study is restricted. In more ambitious settings, such as when a biomarker for a specific cancer subtype is sought, the proteomics scientist might elect to study multiple blood serum samples from multiple cancer patients to minimise confounding factors and account for experimental noise. Thus, complicated experimental designs are sometimes necessary to account for the dynamic complexity of the proteome.

Limitations of Genomics and Proteomics Studies

Proteomics gives a different level of understanding than genomics for many reasons:

- the level of transcription of a gene gives only a rough estimate of its *level of translation* into a protein. An mRNA produced in abundance may be degraded rapidly or translated inefficiently, resulting in a small amount of protein.
- as mentioned above many proteins experience *post-translational modifications* that profoundly affect their activities; for example some proteins are not active until they become phosphorylated. Methods such as phosphoproteomics and glycoproteomics are used to study post-translational modifications.
- many transcripts give rise to more than one protein, through alternative splicing or alternative post-translational modifications.
- many proteins form complexes with other proteins or RNA molecules, and only function in the presence of these other molecules.
- protein degradation rate plays an important role in protein content.

Reproducibility. One major factor affecting reproducibility in proteomics experiments is the simultaneous elution of many more peptides than can be measured by mass spectrometers. This causes stochastic differences between experiments due to data-dependant acquisition of tryptic peptides. Although early large-scale shotgun proteomics analyses showed considerable variability between laboratories, presumably due in part to technical and experimental differences between labs, reproducibility has been improved in more recent mass spectrometry analysis, particularly on the protein level and using Orbitrap mass spectrometers. Notably, targeted proteomics shows increased reproducibility and repeatability compared with shotgun methods, although at the expense of data density and effectiveness.

Methods of Studying Proteins

In proteomics, there are multiple methods to study proteins. Generally, proteins can either be detected using antibodies (immunoassays) or using mass spectrometry. If a complex biological sample is analyzed, then biochemical separation has to be used before the detection step as there are too many analytes in the sample to perform accurate detection and quantification.

Protein Detection with Antibodies (Immunoassays)

Antibodies to particular proteins or to their modified forms have been used in biochemistry and cell biology studies. These are among the most common tools used by molecular biologists today. There are several specific techniques and protocols that use antibodies for protein detection. The enzyme-linked immunosorbent assay (ELISA) has been used for decades to detect and quantitatively measure proteins in samples. The Western blot can be used for detection and quantification of individual proteins, where in an initial step a complex protein mixture is separated using SDS-PAGE and then the protein of interest is identified using an antibody.

Modified proteins can be studied by developing an antibody specific to that modification. For example, there are antibodies that only recognize certain proteins when they are tyrosine-phosphorylated, known as phospho-specific antibodies. Also, there are antibodies specific to other modifications. These can be used to determine the set of proteins that have undergone the modification of interest.

Antibody-free Protein Detection

While protein detection with antibodies are still very common in molecular biology, also other methods have been developed that do not rely on an antibody. These methods offer various advantages, for instance they are often able to determine the sequence of a protein or peptide, they may have higher throughput than antibody-based and they sometimes can identify and quantify proteins for which no antibody exists.

Detection Methods

One of the earliest method for protein analysis has been Edman degradation (introduced in 1967) where a single peptide is subjected to multiple steps of chemical degradation to resolve its sequence. These methods have mostly been supplanted by technologies that offer higher throughput.

More recent methods use mass spectrometry-based techniques, a development that was made possible by the discovery of "soft ionization" methods such as matrix-assisted laser desorption/ionization (MALDI) and electrospray ionization (ESI) developed in the 1980s. These methods gave rise to the top-down and the bottom-up proteomics workflows where often additional separation is performed before analysis.

Separation Methods

For the analysis of complex biological samples, a reduction of sample complexity is required. This can be performed off-line by one-dimensional or two dimensional separation. More recently, on-line methods have been developed where individual peptides (in bottom-up proteomics approaches) are separated using Reversed-phase chromatography and then directly ionized using ESI; the direct coupling of separation and analysis explains the term "on-line" analysis.

Hybrid Technologies

There are several hybrid technologies that use antibody-based purification of individual analytes and then perform mass spectrometric analysis for identification and quantification. Examples of these methods are the MSIA (mass spectrometric immunoassay) developed by Randall Nelson in 1995 and the SISCAPA (Stable Isotope Standard Capture with Anti-Peptide Antibodies) method, introduced by Leigh Anderson in 2004.

Current Research Methodologies

Fluorescence two-dimensional differential gel electrophoresis (2-D DIGE) can be used to quantify variation in the 2-D DIGE process and establish statistically valid thresholds for assigning quantitative changes between samples.

Comparative proteomic analysis can reveal the role of proteins in complex biological systems, including reproduction. For example, treatment with the insecticide triazophos causes an increase in the content of brown planthopper (*Nilaparvata lugens* (Stål)) male accessory gland proteins (Acps) that can be transferred to females via mating, causing an increase in fecundity (i.e. birth rate) of females. To identify changes in the types of accessory gland proteins (Acps) and reproductive proteins that mated female planthoppers received from male planthoppers, researchers conducted a comparative proteomic analysis of mated *N. lugens* females. The results indicated that these proteins participate in the reproductive process of *N. lugens* adult females and males.

Proteome analysis of *Arabidopsis peroxisomes* has been established as the major unbiased approach for identifying new peroxisomal proteins on a large scale.

There are many approaches to characterizing the human proteome, which is estimated to contain between 20,000 and 25,000 non-redundant proteins. The number of unique protein species will likely increase by between 50,000 and 500,000 due to RNA splicing and proteolysis events, and when post-translational modification are also considered, the total number of unique human proteins is estimated to range in the low millions.

In addition, the first promising attempts to decipher the proteome of animal tumors have recently been reported. This method used as a functional method in *Macrobrachium rosenbergii* protein profiling.

High-throughput Proteomic Technologies

Proteomics has steadily gained momentum over the past decade with the evolution of several approaches. Few of these are new and others build on traditional methods. Mass spectrometry-based methods and micro arrays are the most common technologies for large-scale study of proteins.

Mass Spectrometry and Protein Profiling

There are two mass spectrometry-based methods currently used for protein profiling. The more established and widespread method uses high resolution, two-dimensional electrophoresis to separate proteins from different samples in parallel, followed by selection and staining of differentially expressed proteins to be identified by mass spectrometry. Despite the advances in 2DE and its maturity, it has its limits as well. The central concern is the inability to resolve all the proteins within a sample, given their dramatic range in expression level and differing properties.

The second quantitative approach uses stable isotope tags to differentially label proteins from two different complex mixtures. Here, the proteins within a complex mixture are labeled first isotopically, and then digested to yield labeled peptides. The labeled mixtures are then combined, the peptides separated by multidimensional liquid chromatography and analyzed by tandem mass spectrometry. Isotope coded affinity tag (ICAT) reagents are the widely used isotope tags. In this method, the cysteine residues of proteins get covalently attached to the ICAT reagent, thereby reducing the complexity of the mixtures omitting the non-cysteine residues.

Quantitative proteomics using stable isotopic tagging is an increasingly useful tool in modern development. Firstly, chemical reactions have been used to introduce tags into specific sites or proteins for the purpose of probing specific protein functionalities. The isolation of phosphorylated peptides has been achieved using isotopic labeling and selective chemistries to capture the fraction of protein among the complex mixture. Secondly, the ICAT technology was used to differentiate between partially purified or purified macromolecular complexes such as large RNA polymerase II pre-initiation complex and the proteins complexed with yeast transcription factor. Thirdly, ICAT labeling was recently combined with chromatin isolation to identify and quantify chromatin-associated proteins. Finally ICAT reagents are useful for proteomic profiling of cellular organelles and specific cellular fractions.

Another quantitative approach is the Accurate Mass and Time (AMT) tag approach developed by Richard D. Smith and coworkers at Pacific Northwest National Laboratory. In this approach, increased throughput and sensitivity is achieved by avoiding the need for tandem mass spectrometry, and making use of precisely determined separation time information and highly accurate mass determinations for peptide and protein identifications.

Protein Chips

Balancing the use of mass spectrometers in proteomics and in medicine is the use of protein micro arrays. The aim behind protein micro arrays is to print thousands of protein detecting features for the interrogation of biological samples. Antibody arrays are an example in which a host of different antibodies are arrayed to detect their respective antigens from a sample of human blood. Another approach is the arraying of multiple protein types for the study of properties like protein-DNA, protein-protein and protein-ligand interactions. Ideally, the functional proteomic arrays would contain the entire complement of the proteins of a given organism. The first version of such arrays consisted of 5000 purified proteins from yeast deposited onto glass microscopic slides. Despite the success of first chip, it was a greater challenge for protein arrays to be implemented. Proteins are inherently much more difficult to work with than DNA. They have a broad dynamic range, are less stable than DNA and their structure is difficult to preserve on glass slides, though they are essential for most assays. The global ICAT technology has striking advantages over protein chip technologies.

Reverse-phased Protein Microarrays

This is a promising and newer microarray application for the diagnosis, study and treatment of complex diseases such as cancer. The technology merges laser capture microdissection (LCM) with micro array technology, to produce reverse phase protein microarrays. In this type of microarrays, the whole collection of protein themselves are immobilized with the intent of capturing various stages of disease within an individual patient. When used with LCM, reverse phase arrays can monitor the fluctuating state of proteome among different cell population within a small area of human tissue. This is useful for profiling the status of cellular signaling molecules, among a cross section of tissue that includes both normal and cancerous cells. This approach is useful in monitoring the status of key factors in normal prostate epithelium and invasive prostate cancer tissues. LCM then dissects these tissue and protein lysates were arrayed onto nitrocellulose slides, which were probed with specific antibodies. This method can track all kinds of molecular events and can compare diseased and healthy tissues within the same patient enabling the development of treatment strategies and diagnosis. The ability to acquire proteomics snapshots of neighboring cell populations, using reverse phase microarrays in conjunction with LCM has a number of applications beyond the study of tumors. The approach can provide insights into normal physiology and pathology of all the tissues and is invaluable for characterizing developmental processes and anomalies.

Practical Applications of Proteomics

One major development to come from the study of human genes and proteins has been the identification of potential new drugs for the treatment of disease. This relies on genome and proteome information to identify proteins associated with a disease, which computer software can then use as targets for new drugs. For example, if a certain protein is implicated in a disease, its 3D structure provides the information to design drugs to interfere with the action of the protein. A molecule that fits the active site of an enzyme, but cannot be released by the enzyme, inactivates the enzyme. This is the basis of new drug-discovery tools, which aim to find new drugs to inactivate proteins involved in disease. As genetic differences among individuals are found, researchers expect to use these techniques to develop personalized drugs that are more effective for the individual.

Proteomics is also used to reveal complex plant-insect interactions that help identify candidate genes involved in the defensive response of plants to herbivory.

Interaction Proteomics and Protein Networks

Interaction proteomics is the analysis of protein interactions from scales of binary interactions to proteome- or network-wide. Most proteins function via protein-protein interactions, and one goal of interaction proteomics is to identify binary protein interactions, protein complexes, and interactomes.

Several methods are available to probe protein–protein interactions. While the most traditional method is yeast two-hybrid analysis, a powerful emerging method is affinity purification followed by protein mass spectrometry using tagged protein baits. Other methods include surface plasmon resonance (SPR), protein microarrays, dual polarisation interferometry, microscale thermophoresis and experimental methods such as phage display and *in silico* computational methods.

Knowledge of protein-protein interactions is especially useful in regard to biological networks and systems biology, for example in cell signaling cascades and gene regulatory networks (GRNs, where knowledge of protein-DNA interactions is also informative). Proteome-wide analysis of protein interactions, and integration of these interaction patterns into larger biological networks, is crucial towards understanding systems-level biology.

Expression Proteomics

Expression proteomics includes the analysis of protein expression at larger scale. It helps identify main proteins in a particular sample, and those proteins differentially expressed in related samples—such as diseased vs. healthy tissue. If a protein is found only in a diseased sample then it can be a useful drug target or diagnostic marker. Proteins with same or similar expression profiles may also be functionally related. There are technologies such as 2D-PAGE and mass spectrometry that are used in expression proteomics.

Biomarkers

The National Institutes of Health has defined a biomarker as "a characteristic that is objectively measured and evaluated as an indicator of normal biological processes, pathogenic processes, or pharmacologic responses to a therapeutic intervention."

Understanding the proteome, the structure and function of each protein and the complexities of protein–protein interactions is critical for developing the most effective diagnostic techniques and disease treatments in the future. For example, proteomics is highly useful in identification of candidate biomarkers (proteins in body fluids that are of value for diagnosis), identification of the bacterial antigens that are targeted by the immune response, and identification of possible immunohistochemistry markers of infectious or neoplastic diseases.

An interesting use of proteomics is using specific protein biomarkers to diagnose disease. A number of techniques allow to test for proteins produced during a particular disease, which helps to diagnose the disease quickly. Techniques include western blot, immunohistochemical staining, enzyme linked immunosorbent assay (ELISA) or mass spectrometry. Secretomics, a subfield of proteomics that studies secreted proteins and secretion pathways using proteomic approaches, has recently emerged as an important tool for the discovery of biomarkers of disease.

Proteogenomics

In what is now commonly referred to as proteogenomics, proteomic technologies such as mass spectrometry are used for improving gene annotations. Parallel analysis of the genome and the proteome facilitates discovery of post-translational modifications and proteolytic events, especially when comparing multiple species (comparative proteogenomics).

Structural Proteomics

Structural proteomics includes the analysis of protein structures at large-scale. It compares protein structures and helps identify functions of newly discovered genes. The structural analysis also helps to understand that where drugs bind to proteins and also show where proteins interact with each other. This understanding is achieved using different technologies such as X-ray crystallography and NMR spectroscopy.

Bioinformatics for Proteomics (Proteome Informatics)

There is a large amount of proteomics data being collected with the help of high throughput technologies such as mass spectrometry and microarray. It would often take weeks or months to analyze the data and perform comparisons by hand. For this reason, biologists and chemists are collaborating with computer scientists and mathematicians to create programs and pipeline to computationally analyze the protein data. Using bioinformatics techniques, researchers are capable of faster analysis and data storage. A good place to find lists of current programs and databases is on the ExPASy bioinformatics resource portal <http://www.expasy.org/proteomics>. The applications of bioinformatics-based proteomics includes medicine, disease diagnosis, biomarker identification, and many more.

Protein Identification

Mass spectrometry and microarray produce peptide fragmentation information but do not give identification of specific proteins present in the original sample. Due to the lack of specific protein identification, past researchers were forced to decipher the peptide fragments themselves. However, there are currently programs available for protein identification. These programs take the peptide sequences output from mass spectrometry and microarray and return information about matching or similar proteins. This is done through algorithms implemented by the program which perform alignments with proteins from known databases such as UniProt and PROSITE to predict what proteins are in the sample with a degree of certainty.

Protein Structure

The biomolecular structure forms the 3D configuration of the protein. Understanding the protein's structure aids in identification of the protein's interactions and function. It used to be that the 3D structure of proteins could only be determined using X-ray crystallography and NMR spectroscopy. Now, through bioinformatics, there are computer programs that can predict and model the structure of proteins. These programs use the chemical properties of amino acids and structural properties of known proteins to predict the 3D model of sample proteins. This also allows scientists to take a look at protein interactions on a larger scale. In addition, biomedical engineers are developing methods to factor in the flexibility of protein structures to make comparisons and predictions.

Post-translational Modifications

Unfortunately, most programs available for protein analysis are not written for proteins that have undergone post-translational modifications. Some programs will accept post-translational modifications to aid in protein identification but then ignore the modification during further protein analysis. It is important to account for these modifications since they can affect the protein's structure. In turn, computational analysis of post-translational modifications has gained the attention of the scientific community. The current post-translational modification programs are only predictive. Chemists, biologists and computer scientists are working together to create and introduce new pipelines that allow for analysis of post-translational modifications that have been experimentally identified for their effect on the protein's structure and function.

Computational Methods in Studying Protein Biomarkers

One example of the use of bioinformatics and the use of computational methods is the study of protein biomarkers. Computational predictive models have shown that extensive and diverse feto-maternal protein trafficking occurs during pregnancy and can be readily detected non-invasively in maternal whole blood. This computational approach circumvented a major limitation, the abundance of maternal proteins interfering with the detection of fetal proteins, to fetal proteomic analysis of maternal blood. Computational models can use fetal gene transcripts previously identified in maternal whole blood to create a comprehensive proteomic network of the term neonate. Such work shows that the fetal proteins detected in pregnant woman's blood originate from a diverse group of tissues and organs from the developing fetus. The proteomic networks contain many biomarkers that are proxies for development and illustrate the potential clinical application of this technology as a way to monitor normal and abnormal fetal development.

An information theoretic framework has also been introduced for biomarker discovery, integrating biofluid and tissue information. This new approach takes advantage of functional synergy between certain biofluids and tissues with the potential for clinically significant findings not possible if tissues and biofluids were considered individually. By conceptualizing tissue-biofluid as information channels, significant biofluid proxies can be identified and then used for guided development of clinical diagnostics. Candidate biomarkers are then predicted based on information transfer criteria across the tissue-biofluid channels. Significant biofluid-tissue relationships can be used to prioritize clinical validation of biomarkers.

Emerging Trends in Proteomics

A number of emerging concepts have the potential to improve current features of proteomics. Obtaining absolute quantification of proteins and monitoring post-translational modifications are the two tasks that impact the understanding of protein function in healthy and diseased cells. For many cellular events, the protein concentrations do not change; rather, their function is modulated by post-transitional modifications (PTM). Methods of monitoring PTM are an underdeveloped area in proteomics. Selecting a particular subset of protein for analysis substantially reduces protein complexity, making it advantageous for diagnostic purposes where blood is the starting material. Another important aspect of proteomics, yet not addressed, is that proteomics methods should focus on studying proteins in the context of the environment. The increasing use of chemical cross linkers, introduced into living cells to fix protein-protein, protein-DNA and other interactions, may ameliorate this problem partially. The challenge is to identify suitable methods of preserving relevant interactions. Another goal for studying protein is to develop more sophisticated methods to image proteins and other molecules in living cells and real time.

Proteomics for Systems Biology

Advances in quantitative proteomics would clearly enable more in-depth analysis of cellular systems. Biological systems are subject to a variety of perturbations (cell cycle, cellular differentiation, carcinogenesis, environment (biophysical), etc.). Transcriptional and translational responses to these perturbations results in functional changes to the proteome implicated in response to the stimulus. Therefore, describing and quantifying proteome-wide changes in protein abundance is crucial towards understanding biological phenomenon more holistically, on the level of the entire system. In this way, proteomics can be seen as complementary to genomics, transcriptomics, epigenomics, metabolomics, and other -omics approaches in integrative analyses attempting to define biological phenotypes more comprehensively. As an example, *The Cancer Proteome Atlas* provides quantitative protein expression data for ~200 proteins in over 4,000 tumor samples with matched transcriptomic and genomic data from The Cancer Genome Atlas. Similar datasets in other cell types, tissue types, and species, particularly using deep shotgun mass spectrometry, will be an immensely important resource for research in fields like cancer biology, developmental and stem cell biology, medicine, and evolutionary biology.

Human Plasma Proteome

Characterizing the human plasma proteome has become a major goal in the proteomics arena. The plasma proteome represents one of the most complex proteomes in the human body. It contains immunoglobulin, cytokines, protein hormones, and secreted proteins indicative of infection on top of resident, hemostatic proteins. It also contains tissue leakage proteins due to the blood circulation through different tissues in the body. The blood thus contains information on the physiological state of all tissues and, combined with its accessibility, makes the blood proteome invaluable for medical purposes. Even with the recent advancements in proteomics, characterizing the proteome of blood plasma is a daunting challenge.

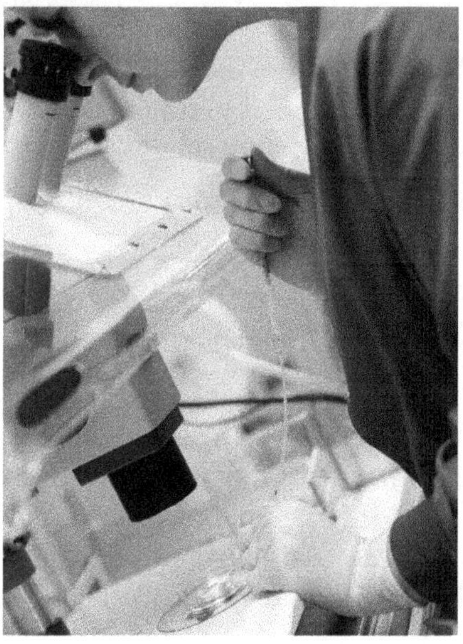

Temporal and spatial dynamics further complicate the study of human plasma proteome. The turnover of some proteins is quite faster than others and the protein content of an artery may substantially vary from that of a vein. All these differences make even the simplest proteomic task of cataloging the proteome seem out of reach. To tackle this problem, priorities need to be established. Capturing the most meaningful subset of proteins among the entire proteome to generate a diagnostic tool is one such priority. Secondly, since cancer is associated with enhanced glycosylation of proteins, methods that focus on this part of proteins will also be useful. Again: multiparameter analysis best reveals a pathological state. As these technologies improve, the disease profiles should be continually related to respective gene expression changes.

Protein–protein Interaction

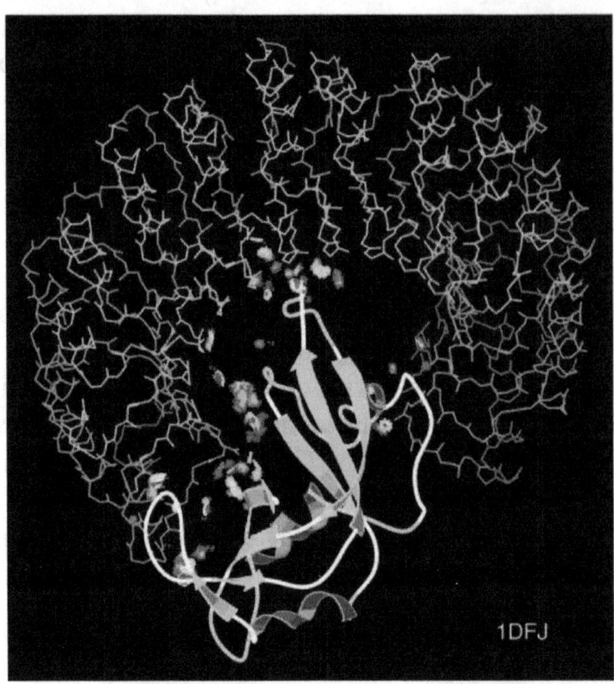

The horseshoe shaped ribonuclease inhibitor (shown as wireframe) forms a protein–protein interaction with the ribonuclease protein. The contacts between the two proteins are shown as coloured patches.

Protein–protein interactions (PPIs) refer to lasting or ephemeral physical contacts of high specificity established between two or more protein molecules as a result of biochemical events steered by electrostatic forces including the hydrophobic effect. Commonly they refer to physical contacts with molecular associations between chains that occur in a cell or in a living organism in a specific biomolecular context.

Proteins rarely act alone as their functions tend to be regulated. Many molecular processes within a cell are carried out by molecular machines that are built from a large number of protein components organized by their PPIs. These interactions make up the so-called interactomics of the organism, while aberrant PPIs are the basis of multiple aggregation-related diseases, such as Creutzfeld-Jacob, Alzheimer's disease, and may lead to cancer.

PPIs have been studied from different perspectives: biochemistry, quantum chemistry, molecular dynamics, signal transduction, among others. All this information enables the creation of large protein interaction networks – similar to metabolic or genetic/epigenetic networks – that empower the current knowledge on biochemical cascades and molecular etiology of disease, as well as the discovery of putative protein targets of therapeutic interest.

Examples

Signal transduction

The activity of the cell is regulated by extracellular signals. Signals propagation to inside and/

or along the interior of cells depends on PPIs between the various signaling molecules. The recruitment of signaling pathways through PPIs is called signal transduction and plays a fundamental role in many biological processes and in many diseases including Parkinson's disease and cancer.

Transport across membranes

A protein may be carrying another protein (for example, from cytoplasm to nucleus or vice versa in the case of the nuclear pore importins).

Cell metabolism

In many biosynthetic processes enzymes interact with each other to produce small compounds or other macromolecules.

Muscle contraction

Physiology of muscle contraction involves several interactions. Myosin filaments act as molecular motors and by binding to actin enables filament sliding. Furthermore, members of the skeletal muscle lipid droplet-associated proteins family associate with other proteins, as activator of adipose triglyceride lipase and its coactivator comparative gene identification-58, to regulate lipolysis in skeletal muscle.

Types

To describe the types of protein–protein interactions (PPIs) it is important to consider that proteins can interact in a "transient" way (to produce some specific effect in a short time) or to interact with other proteins in a "stable" way to build multiprotein complexes that are molecular machines within the living systems. A protein complex assembly can result in the formation of homo-oligomeric or hetero-oligomeric complexes. In addition to the conventional complexes, as enzyme-inhibitor and antibody-antigen, interactions can also be established between domain-domain and domain-peptide. Another important distinction to identify protein-protein interactions is the way they have been determined, since there are techniques that measure direct physical interactions between protein pairs, named "binary" methods, while there are other techniques that measure physical interactions among groups of proteins, without pairwise determination of protein partners, named "co-complex" methods.

Homo-oligomers Vs. Hetero-oligomers

Homo-oligomers are macromolecular complexes constituted by only one type of protein subunit. Protein subunits assembly is guided by the establishment of non-covalent interactions in the quaternary structure of the protein. Disruption of homo-oligomers in order to return to the initial individual monomers often requires denaturation of the complex. Several enzymes, carrier proteins, scaffolding proteins, and transcriptional regulatory factors carry out their functions as homo-oligomers. Distinct protein subunits interact in hetero-oligomers, which are essential to control several cellular functions. The importance of the communication between heterologous proteins is even more evident during cell signaling events and such interactions are only possible due to structural domains within the proteins (as described below).

Stable Interactions Vs. Transient Interactions

Stable interactions involve proteins that interact for a long time, taking part of permanent complexes as subunits, in order to carry out structural or functional roles. These are usually the case of homo-oligomers (e.g. cytochrome c), and some hetero-oligomeric proteins, as the subunits of ATPase. On the other hand, a protein may interact briefly and in a reversible manner with other proteins in only certain cellular contexts – cell type, cell cycle stage, external factors, presence of other binding proteins, etc. – as it happens with most of the proteins involved in biochemical cascades. These are called transient interactions . For example, some G protein-coupled receptors only transiently bind to $G_{i/o}$ proteins when they are activated by extracellular ligands, while some G_q-coupled receptors, such as muscarinic receptor M3, pre-couple with G_q proteins prior to the receptor-ligand binding. Interactions between intrinsically disordered protein regions to globular protein domains (i.e. MoRFs) are transient interactions.

Covalent Vs. Non-Covalent

Covalent interactions are those with the strongest association and are formed by disulphide bonds or electron sharing. Although being rare, these interactions are determinant in some posttranslational modifications, as ubiquitination and SUMOylation. Non-covalent bonds are usually established during transient interactions by the combination of weaker bonds, such as hydrogen bonds, ionic interactions, Van der Waals forces, or hydrophobic bonds.

Role of Water

Water molecules play a significant role in the interactions between proteins. The crystal structures of complexes, obtained at high resolution from different but homologous proteins, have shown that some interface water molecules are conserved between homologous complexes. The majority of the interface water molecules make hydrogen bonds with both partners of each complex. Some interface amino acid residues or atomic groups of one protein partner engage in both direct and water mediated interactions with the other protein partner. Doubly indirect interactions, mediated by two water molecules, are more numerous in the homologous complexes of low affinity. Carefully conducted mutagenesis experiments, e.g. changing a tyrosine residue into a phenylalanine, have shown that water mediated interactions can contribute to the energy of interaction. Thus, water molecules may facilitate the interactions and cross-recognitions between proteins.

Structure

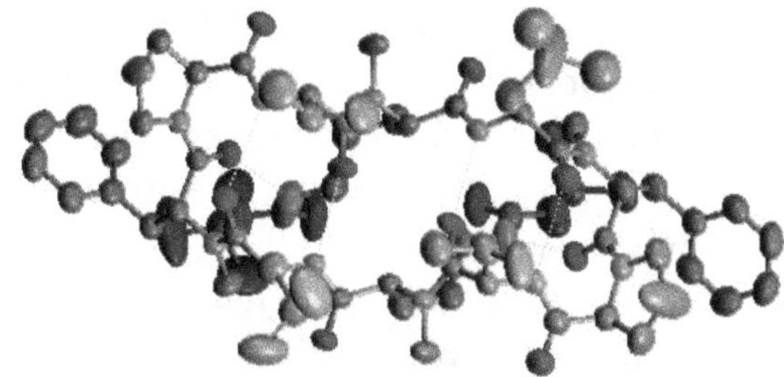

Crystal structure of modified Gramicidin S horizontally determined by X-ray crystallography

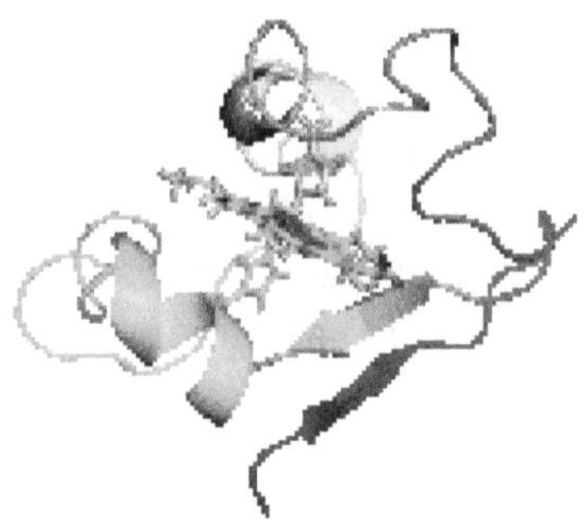

NMR structure of cytochrome C illustrating its dynamics in solution

The molecular structures of many protein complexes have been unlocked by the technique of X-ray crystallography. The first structure to be solved by this method was that of sperm whale myoglobin by Sir John Cowdery Kendrew. In this technique the angles and intensities of a beam of X-rays diffracted by crystalline atoms are detected in a film, thus producing a three-dimensional picture of the density of electrons within the crystal.

Later, nuclear magnetic resonance also started to be applied with the aim of unravelling the molecular structure of protein complexes. One of the first examples was the structure of calmodulin-binding domains bound to calmodulin. This technique is based on the study of magnetic properties of atomic nuclei, thus determining physical and chemical properties of the correspondent atoms or the molecules. Nuclear magnetic resonance is advantageous for characterizing weak PPIs.

Domains

Proteins hold structural domains that allow their interaction with and bind to specific sequences on other proteins:

- *Src homology 2 (SH2) domain*

SH2 domains are structurally composed by three-stranded twisted beta sheet sandwiched flanked by two alpha-helices. The existence of a deep binding pocket with high affinity for phosphotyrosine, but not for phosphoserine or phosphothreonine, is essential for the recognition of tyrosine phosphorylated proteins, mainly autophosphorylated growth factor receptors. Growth factor receptor binding proteins and phospholipase Cγ are examples of proteins that have SH2 domains.

- *Src homology 3 (SH3) domain*

Structurally, SH3 domains are constituted by a beta barrel formed by two orthogonal beta sheets and three anti-parallel beta strands. These domains recognize proline enriched sequences, as polyproline type II helical structure (PXXP motifs) in cell signaling proteins like protein tyrosine kinases and the growth factor receptor bound protein 2 (Grb2).

- *Phosphotyrosine-binding (PTB) domain*

PTB domains interact with sequences that contain a phosphotyrosine group. These domains can be found in the insulin receptor substrate.

- *LIM domain*

LIM domains were initially identified in three homeodomain transcription factors (lin11, is11, and mec3). In addition to this homeodomain proteins and other proteins involved in development, LIM domains have also been identified in non-homeodomain proteins with relevant roles in cellular differentiation, association with cytoskeleton and senescence. These domains contain a tandem cysteine-rich Zn^{2+}-finger motif and embrace the consensus sequence CX2CX16-23HX2CX2CX-2CX16-21CX2C/H/D. LIM domains bind to PDZ domains, bHLH transcription factors, and other LIM domains.

- *Sterile alpha motif (SAM) domain*

SAM domains are composed by five helices forming a compact package with a conserved hydrophobic core. These domains, which can be found in the Eph receptor and the stromal interaction molecule (STIM) for example, bind to non-SAM domain-containing proteins and they also appear to have the ability to bind RNA.

- *PDZ domain*

PDZ domains were first identified in three guanylate kinases: PSD-95, DlgA and ZO-1. These domains recognize carboxy-terminal tri-peptide motifs (S/TXV), other PDZ domains or LIM domains and bind them through a short peptide sequence that has a C-terminal hydrophobic residue. Some of the proteins identified as having PDZ domains are scaffolding proteins or seem to be involved in ion receptor assembling and receptor-enzyme complexes formation.

- *FERM domain*

FERM domains contain basic residues capable of binding PtdIns(4,5)P_2. Talin and focal adhesion kinase (FAK) are two of the proteins that present FERM domains.

- *Calponin homology (CH) domain*

CH domains are mainly present in cytoskeletal proteins as parvin.

- *Pleckstrin homology domain*

Pleckstrin homology domains bind to phosphoinositides and acid domains in signaling proteins.

- *WW domain*

WW domains bind to proline enriched sequences.

- *WSxWS motif*

Found in cytokine receptors

Properties of the Interface

The study of the molecular structure can give fine details about the interface that enables the interaction between proteins. When characterizing PPI interfaces it is important to take into account the type of complex.

Parameters evaluated include size (measured in absolute dimensions Å^2 or in solvent-accessible surface area (SASA)), shape, complementarity between surfaces, residue interface propensities, hydrophobicity, segmentation and secondary structure, and conformational changes on complex formation.

The great majority of PPI interfaces reflects the composition of protein surfaces, rather than the protein cores, in spite of being frequently enriched in hydrophobic residues, particularly in aromatic residues. PPI interfaces are dynamic and frequently planar, although they can be globular and protruding as well. Based on three structures – insulin dimer, trypsin-pancreatic trypsin inhibitor complex, and oxyhaemoglobin – Cyrus Chothia and Joel Janin found that between 1,130 and 1,720 Å² of surface area was removed from contact with water indicating that hydrophobicity is a major factor of stabilization of PPIs. Later studies refined the buried surface area of the majority of interactions to 1,600±350 Å². However, much larger interaction interfaces were also observed and were associated with significant changes in conformation of one of the interaction partners. PPIs interfaces exhibit both shape and electrostatic complementarity.

Regulation

- Protein concentration, which in turn are affected by expression levels and degradation rates;
- Protein affinity for proteins or other binding ligands;
- Ligands concentrations (substrates, ions, etc.);
- Presence of other proteins, nucleic acids, and ions;
- Electric fields around proteins.
- Occurrence of covalent modifications;

Measurement

There are a multitude of methods to detect them. Each of the approaches has its own strengths and weaknesses, especially with regard to the sensitivity and specificity of the method. The most conventional and widely used high-throughput methods are yeast two-hybrid screening and affinity purification coupled to mass spectrometry.

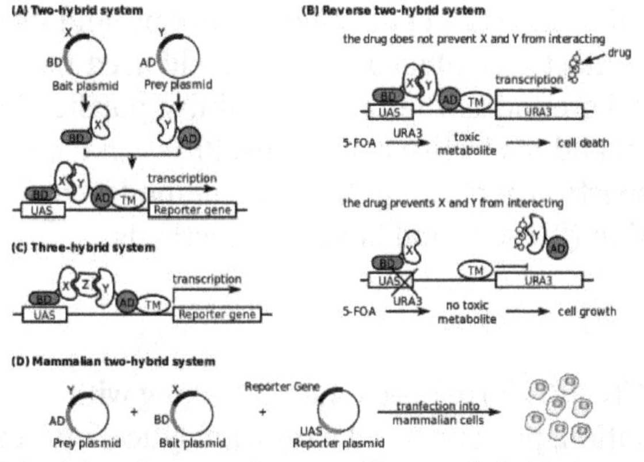

Principles of yeast and mammalian two-hybrid systems

Yeast Two-hybrid Screening

This system was firstly described in 1989 by Fields and Song using *Saccharomyces cerevisiae* as biological model. Yeast two hybrid allows the identification of pairwise PPIs (binary method) *in vivo*, indicating non-specific tendencies towards sticky interactions.

Yeast cells are transfected with two plasmids: the bait (protein of interest fused with the DNA-binding domain of a yeast transcription factor, like Gal4), and the prey (a library of cDNA fragments linked to the activation domain of the transcription factor. Transcription of reporter genes does not occur unless bait and prey interact with each other and form a functional transcription factor. Thus, the interaction between proteins can be inferred by the presence of the products resultant of the reporter gene expression.

Despite its usefulness, the yeast two-hybrid system has limitations: specificity is relatively low; uses yeast as main host system, which can be a problem when studying other biological models; the number of PPIs identified is usually low because some transient PPIs are lost during purification steps; and, understates membrane proteins, for example. Limitations have been overcoming by the emergence of yeast two-hybrid variants, such as the membrane yeast two-hybrid (MYTH) and the split-ubiquitin system, which are not limited to interactions that occur in the nucleus; and, the bacterial two-hybrid system, performed in bacteria;

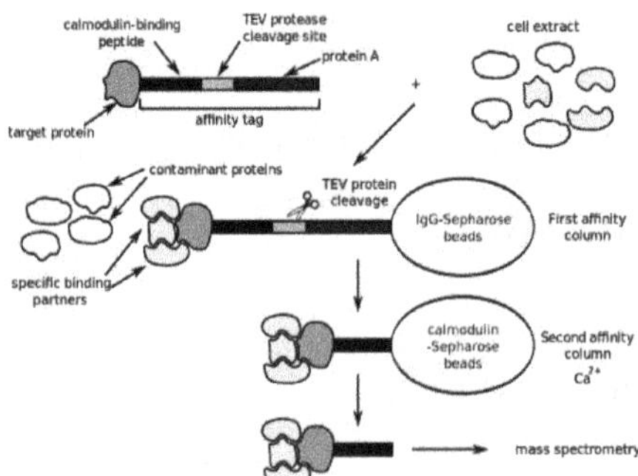

Principle of Tandem Affinity Purification

Affinity Purification Coupled to Mass Spectrometry

Affinity purification coupled to mass spectrometry mostly detects stable interactions and thus better indicates functional in vivo PPIs. This method starts by purification of the tagged protein, which is expressed in the cell usually at *in vivo* concentrations, and its interacting proteins (affinity purification). One of the most advantageous and widely used method to purify proteins with very low contaminating background is the tandem affinity purification, developed by Bertrand Seraphin and Mathias Mann and respective colleagues. PPIs can then be quantitatively and qualitatively analysed by mass spectrometry using different methods: chemical incorporation, biological or metabolic incorporation (SILAC), and label-free methods.

Other Potential Methods

Diverse techniques to identify PPIs have been emerging along with technology progression. These include co-immunoprecipitation, protein microarrays, analytical ultracentrifugation, light scattering, fluorescence spectroscopy, luminescence-based mammalian interactome mapping (LUMIER), resonance-energy transfer systems, mammalian protein–protein interaction trap, electro-switchable biosurfaces, protein-fragment complementation assay, as well as real-time label-free measurements by surface plasmon resonance, and calorimetry.

Text Mining Methods

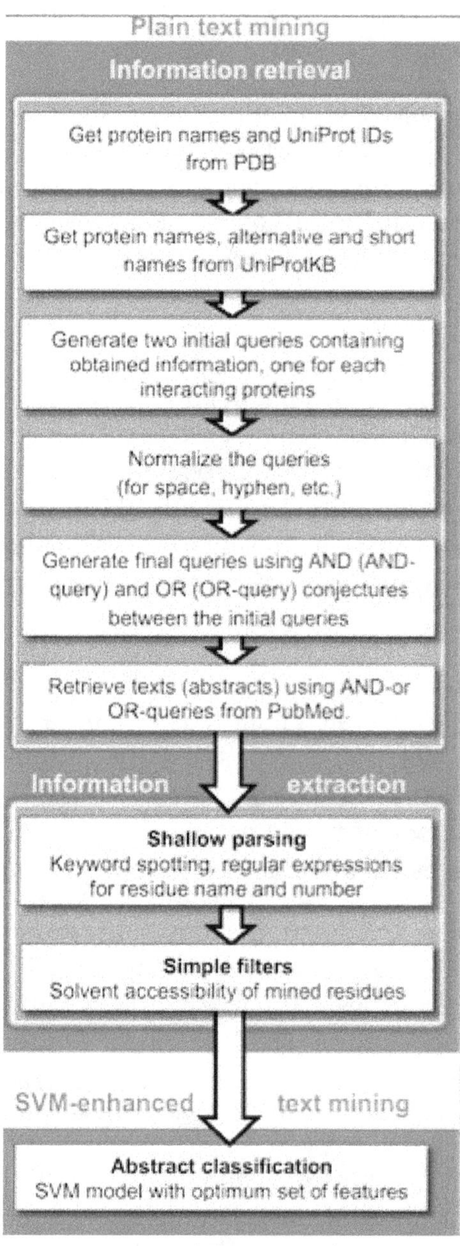

Text mining protocol.

Publicly available information from biomedical research is readily accessible through the internet and is becoming a powerful resource for predictive protein-protein interactions and protein docking. Text mining is much less time costly and consuming compared to other high-throughput techniques. Currently, these methods generally detect binary relations between interacting protein from individual sentences using machine learning and rule/pattern-based information extraction and machine learning approaches. A wide variety of text mining predicting PPIs applications are available for public use, as well as repositories which often stores manually validated and/or computationally predicted PPIs. The principal stages of text mining divides the implementation into two stages: *information retrieval*, where literature abstracts containing names of either or both proteins complexes are selected and *information extraction,* where detecting occurrences of residues are retrieved. The extraction is automated by searching for co-existing sentences, abstracts or paragraphs within textual context.

There are also studies using phylogenetic profiling, basing their functionalities on the theory that proteins involved in common pathways co-evolve in a correlated fashion across large number of species. More complex text mining methodologies use advanced dictionaries and generate networks by Natural Language Processing (NLP) of text, considering gene names as nodes and verbs as edges, other developments involve kernel methods to predict protein interactions.

Machine Learning Methods

These methods use machine learning to distinguish how interacting protein pairs differ from non-interacting protein pairs in terms of pairwise features such as cellular colocalization, gene co-expression, how closely located on a DNA are the genes that encode the two proteins, and so on. Random Forest has been found to be most-effective machine learning method for protein interaction prediction. Such methods have been applied for discovering protein interactions on human interactome, specifically the interactome of Membrane proteins and the interactome of Schizophrenia-associated proteins.

Databases

Large scale identification of PPIs generated hundreds of thousands interactions, which were collected together in specialized biological databases that are continuously updated in order to provide complete interactomes. The first of these databases was the Database of Interacting Proteins (DIP). Since that time, the number of public databases has been increasing. Databases can be subdivided into primary databases, meta-databases, and prediction databases.

- *Primary databases* collect information about published PPIs proven to exist via small-scale or large-scale experimental methods. Examples: DIP, Biomolecular Interaction Network Database (BIND), Biological General Repository for Interaction Datasets (BioGRID), Human Protein Reference Database (HPRD), IntAct Molecular Interaction Database, Molecular Interactions Database (MINT), MIPS Protein Interaction Resource on Yeast (MIPS-MPact), and MIPS Mammalian Protein–Protein Interaction Database (MIPS-MPPI).
- *Meta-databases* normally result from the integration of primary databases information, but can also collect some original data. Examples: Agile Protein Interactomes Dataserver (APID), The Microbial Protein Interaction Database (MPIDB), and Protein Interaction Network Analysis (PINA) platform., (GPS-Prot)
- *Prediction databases* include many PPIs that are predicted using several techniques. Examples: Human Protein–Protein Interaction Prediction Database (PIPs), Interlogous Interaction Database (I2D), Known and Predicted Protein–Protein Interactions (STRING), and Unified Human Interactive (UniHI).

Interaction Networks

Schziophrenia PPI.

Information found in PPIs databases supports the construction of interaction networks. Although the PPI network of a given query protein can be represented in textbooks, diagrams of whole cell PPIs are frankly complex and difficult to generate.

One example of a manually produced molecular interaction map is the Kurt Kohn's 1999 map of cell cycle control. Drawing on Kohn's map, Schwikowski et al. in 2000 published a paper on PPIs in yeast, linking together 1,548 interacting proteins determined by two-hybrid screening. They used a layered graph drawing method to find an initial placement of the nodes and then improved the layout using a force-based algorithm.

Bioinformatic tools have been developed to simplify the difficult task of visualizing molecular interaction networks and complement them with other types of data. For instance, Cytoscape is an open-source software widely used and lots of plugins are currently available. Pajek software is advantageous for the visualization and analysis of very large networks.

Identification of functional modules in PPI networks is an important challenge in bioinformatics. Functional modules means a set of proteins that are highly connected to each other in PPI network. It is almost similar problem as community detection in social networks. There are some methods such as Jactive modules and MoBaS. Jactive modules integrate PPI network and gene expression data where as MoBaS integrate PPI network and Genome Wide association Studies.

The awareness of the major roles of PPIs in numerous physiological and pathological processes has been driving the challenge of unravel many interactomes. Examples of published interactomes are the thyroid specific DREAM interactome and the PP1α interactome in human brain.

Signed Interaction Networks

Protein–protein interactions often result in one of the interacting proteins either being 'activated' or 'repressed'. Such effects can be indicated in a PPI network by "signs" (e.g. "activation" or "inhibition"). Although such attributes have been added to networks for a long time, Vinayagam et al.

(2014) coined the term *Signed network* for them. Signed networks are often expressed by labeling the interaction as either positive or negative. A positive interaction is one where the interaction results in one of the proteins being activated. Conversely a negative interaction indicates that one of the proteins being inactivated.

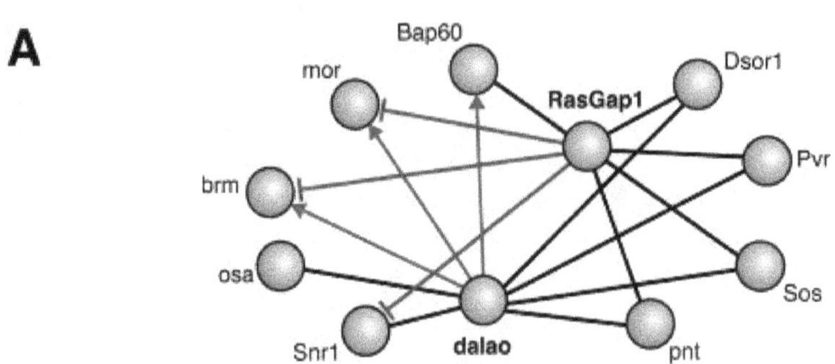

The protein protein interactions are displayed in a signed network that describes what type of interactions that are taking place

Protein–protein interaction networks are often constructed as a result of lab experiments such as yeast two hybrid screens or 'affinity purification and subsequent mass spectrometry techniques. However these methods do not provide the layer of information needed in order to determine what type of interaction is present in order to be able to attribute signs to the network diagrams.

RNA Interference Screens

RNA interference (RNAi) screens (repression of individual proteins between transcription and translation) are one method that can be utilized in the process of providing signs to the protein-protein interactions. Individual proteins are repressed and the resulting phenotypes are analyzed. A correlating phenotypic relationship (i.e. where the inhibition of either of two proteins results in the same phenotype) indicates a positive, or activating relationship. Phenotypes that do not correlate (i.e. where the inhibition of either of two proteins results in two different phenotypes) indicate a negative or inactivating relationship. If protein A is dependent on protein B for activation then the inhibition of either protein A or B will result in a cell losing the service that is provided by protein A and the phenotypes will be the same for the inhibition of either A or B. If, however, protein A is inactivated by protein B then the phenotypes will differ depending on which protein is inhibited (inhibit protein B and it can no longer inactivate protein A leaving A active however inactivate A and there is nothing for B to activate since A is inactive and the phenotype changes). Multiple RNAi screens need to be performed in order to reliably appoint a sign to a given protein-protein interaction. Vinayagam et al. who devised this technique state that a minimum of nine RNAi screens are required with confidence increasing as one carries out more screens.

As Therapeutic Targets

Modulation of PPI is challenging and is receiving increasing attention by the scientific community. Several properties of PPI such as allosteric sites and hotspots, have been incorporated into drug-design strategies. The relevance of PPI as putative therapeutic targets for the development of new treatments is particularly evident in cancer, with several ongoing clinical trials within this area. The consensus among these promising targets is, nonetheless, denoted in the already available drugs on the market to treat a multitude of diseases. Examples are Titrobifan, inhibitor of the glycoprotein IIb/IIIa, used as a cardiovascular drug, and Maraviroc, inhibitor of the CCR5-gp120 interaction, used as anti-HIV drug. Recently, Amit Jaiswal and others were able to develop 30 peptides using protein–protein interaction studies to inhibit telomerase recruitment towards telomeres.

Protein–protein Interaction Prediction

Protein–protein interaction prediction is a field combining bioinformatics and structural biology in an attempt to identify and catalog physical interactions between pairs or groups of proteins. Understanding protein–protein interactions is important for the investigation of intracellular signaling pathways, modelling of protein complex structures and for gaining insights into various biochemical processes. Experimentally, physical interactions between pairs of proteins can be inferred from a variety of experimental techniques, including yeast two-hybrid systems, protein-fragment complementation assays (PCA), affinity purification/ mass spectrometry, protein microarrays, fluorescence resonance energy transfer (FRET), and Microscale Thermophoresis (MST). Efforts to experimentally determine the interactome of numerous species are ongoing, and a number of computational methods for interaction prediction have been developed in recent years.

Methods

Proteins that interact are more likely to co-evolve, therefore, it is possible to make inferences about interactions between pairs of proteins based on their phylogenetic distances. It has also been observed in some cases that pairs of interacting proteins have fused orthologues in other organisms. In addition, a number of bound protein complexes have been structurally solved and can be used to identify the residues that mediate the interaction so that similar motifs can be located in other organisms.

Phylogenetic Profiling

Phylogenetic profiling finds pairs of protein families with similar patterns of presence or absence across large numbers of species. This method is based on the hypothesis that potentially interacting proteins should co-evolve and should have orthologs in closely related species. That is, proteins that form complexes or are part of a pathway should be present simultaneously in order for them to function. A phylogenetic profile is constructed for each protein under investigation. The profile is basically a record of whether the protein is present in certain genomes. If two proteins are found to be present and absent in the same genomes, those proteins are deemed likely to be functionally related. A similar method can be applied to protein domains, where profiles are constructed for domains to determine if there are domain interactions. Some drawbacks with the phylogenetic profile methods are that they are computationally expensive to perform, they rely on homology detection between distant organisms, and they only identify if the proteins being investigated are functionally related (part of complex or in same pathway) and not if they have direct interactions.

Prediction of Co-evolved Protein Pairs Based on Similar Phylogenetic Trees

It was observed that the phylogenetic trees of ligands and receptors were often more similar than due to random chance. This is likely because they faced similar selection pressures and co-evolved. This method uses the phylogenetic trees of protein pairs to determine if interactions exist. To do this, homologs of the proteins of interest are found (using a sequence search tool such as BLAST) and multiple-sequence alignments are done (with alignment tools such as Clustal) to build distance matrices for each of the proteins of interest. The distance matrices should then be used to build phylogenetic trees. However, comparisons between phylogenetic trees are difficult, and current methods circumvent this by simply comparing distance matrices. The distance matrices of the proteins are used to calculate a correlation coefficient, in which a larger value corresponds to co-evolution. The benefit of comparing distance matrices instead of phylogenetic trees is that the results do not depend on the method of tree building that was used. The downside is that difference matrices are not perfect representations of phylogenetic trees, and inaccuracies may result from using such a shortcut. Another factor worthy of note is that there are background similarities between the phylogenetic trees of any protein, even ones that do not interact. If left unaccounted for, this could lead to a high false-positive rate. For this reason, certain methods construct a background tree using 16S rRNA sequences which they use as the canonical tree of life. The distance matrix constructed from this tree of life is then subtracted from the distance matrices of the proteins of interest. However, because RNA distance matrices and DNA distance matrices have different scale, presumably because RNA and DNA have different mutation rates, the RNA matrix needs to be rescaled before it can be subtracted from the DNA matrices. By using molecular clock proteins, the scaling coefficient for protein distance/RNA distance can be calculated. This coefficient is used to rescale the RNA matrix.

Rosetta Stone (Gene Fusion) Method

A Rosetta stone protein is a protein chain composed of two fused proteins. It is observed that proteins or domains that interact with one another tend to have homologs in other genomes that are fused into a Rosetta stone protein , such as might arise by gene fusion when two previously separate genes form a new composite one. This evolutionary mechanism can be used to predict protein interactions. If two proteins are separate in one organism but fused in the other, then it is very likely that they will interact in the case where they are expressed as two separate products. The STRING database makes use of this to predict protein-protein interactions. Gene fusion has been extensively studied and large amounts of data are available. Nonetheless, like phylogenetic profile methods, the Rosetta stone method does not necessarily find interacting proteins, as there can be other reasons for the fusion of two proteins, such as optimizing co-expression of the proteins. The most obvious drawback of this method is that there are many protein interactions that cannot be discovered this way; it relies on the presence of Rosetta stone proteins.

Classification Methods

Classification methods use data to train a program (classifier) to distinguish positive examples of interacting protein/domain pairs with negative examples of non-interacting pairs. Popular classifiers used are Random Forest Decision (RFD) and Support Vector Machines. RFD produces results based on the domain composition of interacting and non-interacting protein pairs. When given a protein pair to classify, RFD first creates a representation of the protein pair in a vector. The vector contains all the domain types used to train RFD, and for each domain type the vector also contains a value of 0, 1, or 2. If the protein pair does not contain a certain domain, then the value for that domain is 0. If one of the proteins of the pair contains the domain, then the value is 1. If both proteins contain the domain, then the value is 2. Using training data, RFD constructs a decision forest, consisting of many decision trees. Each decision tree evaluates several domains, and based on the presence or absence of interactions in these domains, makes a decision as to if the protein pair interacts. The vector representation of the protein pair is evaluated by each tree to determine if they are an interacting pair or a non-interacting pair. The forest tallies up all the input from the trees to come up with a final decision. The strength of this method is that it does not assume that domains interact independent of each other. This makes it so that multiple domains in proteins can be used in the prediction. This is a big step up from previous methods which could only predict based on a single domain pair. The limitation of this method is that it relies on the training dataset to produce results. Thus, usage of different training datasets could influence the results.

Inference of Interactions from Homologous Structures

This group of methods makes use of known protein complex structures to predict and structurally model interactions between query protein sequences. The prediction process generally starts by employing a sequence based method (e.g. Interolog) to search for protein complex structures that are homologous to the query sequences. These known complex structures are then used as templates to structurally model the interaction between query sequences. This method has the advantage of not only inferring protein interactions but also suggests models of how proteins interact structurally, which can provide some insights into the atomic level mechanism of that interaction. On the other hand, the ability for these methods to make a prediction is constrained by a limited number of known protein complex structures.

Association Methods

Association methods look for characteristic sequences or motifs that can help distinguish between interacting and non-interacting pairs. A classifier is trained by looking for sequence-signature pairs where one protein contains one sequence-signature, and its interacting partner contains another sequence-signature. They look specifically for sequence-signatures that are found together more often than by chance. This uses a log-odds score which is computed as log2(Pij/PiPj), where Pij is the observed frequency of domains i and j occurring in one protein pair; Pi and Pj are the background frequencies of domains i and j in the data. Predicted domain interactions are those with positive log-odds scores and also having several occurrences within the database. The downside with this method is that it looks at each pair of interacting domains separately, and it assumes that they interact independently of each other.

Identification of Structural Patterns

This method builds a library of known protein–protein interfaces from the PDB, where the interfaces are defined as pairs of polypeptide fragments that are below a threshold slightly larger than the Van der Waals radius of the atoms involved. The sequences in the library are then clustered based on structural alignment and redundant sequences are eliminated. The residues that have a high (generally >50%) level of frequency for a given position are considered hotspots. This library is then used to identify potential interactions between pairs of targets, providing that they have a known structure (i.e. present in the PDB).

Bayesian Network Modelling

Bayesian methods integrate data from a wide variety of sources, including both experimental results and prior computational predictions, and use these features to assess the likelihood that a particular potential protein interaction is a true positive result. These methods are useful because experimental procedures, particularly the yeast two-hybrid experiments, are extremely noisy and produce many false positives, while the previously mentioned computational methods can only provide circumstantial evidence that a particular pair of proteins might interact.

Domain-pair Exclusion Analysis

The domain-pair exclusion analysis detects specific domain interactions that are hard to detect using Bayesian methods. Bayesian methods are good at detecting nonspecific promiscuous interactions and not very good at detecting rare specific interactions. The domain-pair exclusion analysis method calculates an E-score which measures if two domains interact. It is calculated as log(probability that the two proteins interact given that the domains interact/probability that the two proteins interact given that the domains don't interact). The probabilities required in the formula are calculated using an Expectation Maximization procedure, which is a method for estimating parameters in statistical models. High E-scores indicate that the two domains are likely to interact, while low scores indicate that other domains form the protein pair are more likely to be responsible for the interaction. The drawback with this method is that it does not take into account false positives and false negatives in the experimental data.

Supervised Learning Problem

The problem of PPI prediction can be framed as a supervised learning problem. In this paradigm the known protein interactions supervise the estimation of a function that can predict whether an interaction exists or not between two proteins given data about the proteins (e.g., expression levels of each gene in different experimental conditions, location information, phylogenetic profile, etc.).

Relationship to Docking Methods

The field of protein–protein interaction prediction is closely related to the field of protein–protein docking, which attempts to use geometric and steric considerations to fit two proteins of known structure into a bound complex. This is a useful mode of inquiry in cases where both proteins in the pair have known structures and are known (or at least strongly suspected) to interact, but since so many proteins do not have experimentally determined structures, sequence-based interaction prediction methods are especially useful in conjunction with experimental studies of an organism's interactome.

2

Introduction to Protein

Proteins are large molecules consisting of one or more chains of amino acid residues. Some of the aspects elucidated in this chapter are protein domain, protein biosynthesis, protein structure, protein structure prediction and enzymes. This is an introductory chapter that will introduce briefly all the significant aspects of proteins.

Protein

Proteins are large biomolecules, or macromolecules, consisting of one or more long chains of amino acid residues. Proteins perform a vast array of functions within organisms, including catalysing metabolic reactions, DNA replication, responding to stimuli, and transporting molecules from one location to another. Proteins differ from one another primarily in their sequence of amino acids, which is dictated by the nucleotide sequence of their genes, and which usually results in protein folding into a specific three-dimensional structure that determines its activity.

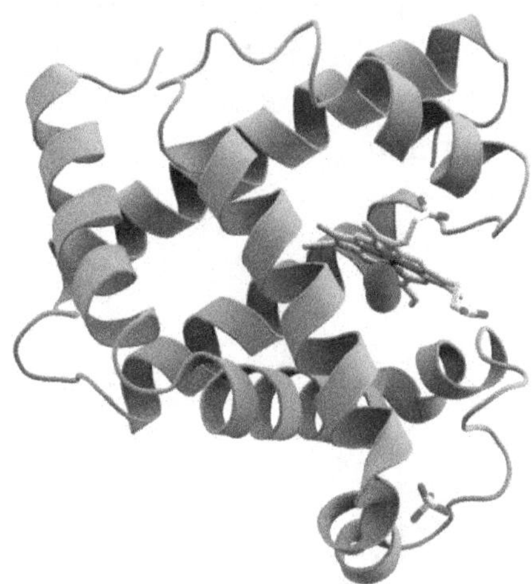

A representation of the 3D structure of the protein myoglobin showing turquoise α-helices. This protein was the first to have its structure solved by X-ray crystallography. Towards the right-center among the coils, a prosthetic group called a heme group (shown in gray) with a bound oxygen molecule (red).

A linear chain of amino acid residues is called a polypeptide. A protein contains at least one long polypeptide. Short polypeptides, containing less than 20–30 residues, are rarely considered to be proteins and are commonly called peptides, or sometimes oligopeptides. The individual amino acid residues are bonded together by peptide bonds and adjacent amino acid residues. The sequence of amino acid residues in a protein is defined by the sequence of a gene, which is encoded in the genetic code. In general, the genetic code specifies 20 standard amino acids; however, in certain organisms the genetic code can include selenocysteine and—in certain archaea—pyrrolysine. Shortly after or even during synthesis, the residues in a protein are often chemically modified by post-translational modification, which alters the physical and chemical properties, folding, stability, activity, and ultimately, the function of the proteins. Sometimes proteins have non-peptide groups attached, which can be called prosthetic groups or cofactors. Proteins can also work together to achieve a particular function, and they often associate to form stable protein complexes.

Once formed, proteins only exist for a certain period of time and are then degraded and recycled by the cell's machinery through the process of protein turnover. A protein's lifespan is measured in terms of its half-life and covers a wide range. They can exist for minutes or years with an average lifespan of 1–2 days in mammalian cells. Abnormal and or misfolded proteins are degraded more rapidly either due to being targeted for destruction or due to being unstable.

Like other biological macromolecules such as polysaccharides and nucleic acids, proteins are essential parts of organisms and participate in virtually every process within cells. Many proteins are enzymes that catalyse biochemical reactions and are vital to metabolism. Proteins also have structural or mechanical functions, such as actin and myosin in muscle and the proteins in the cytoskeleton, which form a system of scaffolding that maintains cell shape. Other proteins are important in cell signaling, immune responses, cell adhesion, and the cell cycle. In animals, proteins are needed in the diet to provide the essential amino acids that cannot be synthesized. Digestion breaks the proteins down for use in the metabolism.

Proteins may be purified from other cellular components using a variety of techniques such as ultracentrifugation, precipitation, electrophoresis, and chromatography; the advent of genetic engineering has made possible a number of methods to facilitate purification. Methods commonly used to study protein structure and function include immunohistochemistry, site-directed mutagenesis, X-ray crystallography, nuclear magnetic resonance and mass spectrometry.

Biochemistry

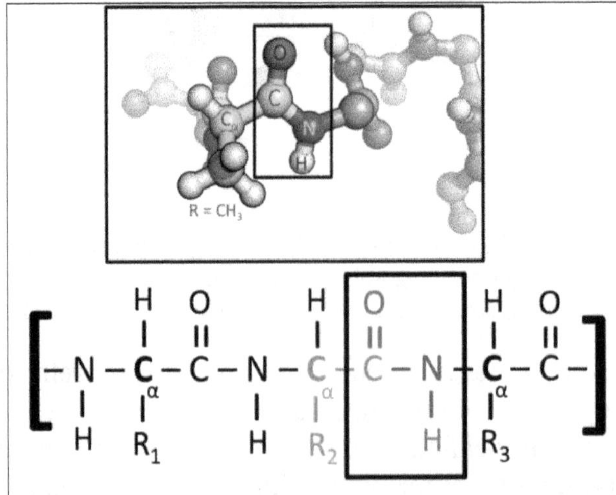

Chemical structure of the peptide bond (bottom) and the three-dimensional structure of a peptide bond between an alanine and an adjacent amino acid (top/inset)

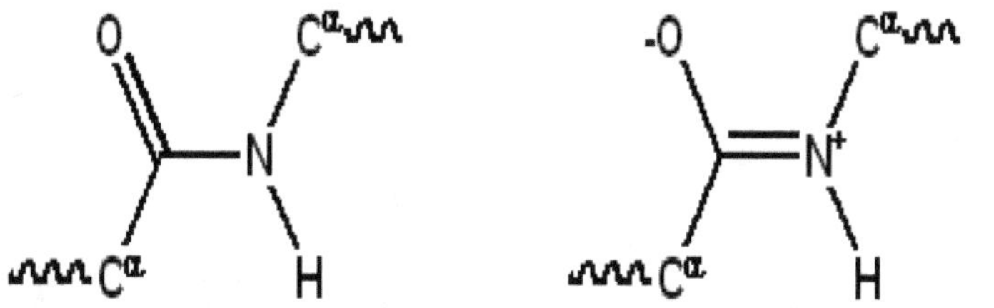

Resonance structures of the peptide bond that links individual amino acids to form a protein polymer

Most proteins consist of linear polymers built from series of up to 20 different L-α-amino acids. All proteinogenic amino acids possess common structural features, including an α-carbon to which an amino group, a carboxyl group, and a variable side chain are bonded. Only proline differs from this basic structure as it contains an unusual ring to the N-end amine group, which forces the CO–NH amide moiety into a fixed conformation. The side chains of the standard amino acids, detailed in the list of standard amino acids, have a great variety of chemical structures and properties; it is the combined effect of all of the amino acid side chains in a protein that ultimately determines its three-dimensional structure and its chemical reactivity. The amino acids in a polypeptide chain are linked by peptide bonds. Once linked in the protein chain, an individual amino acid is called a *residue,* and the linked series of carbon, nitrogen, and oxygen atoms are known as the *main chain* or *protein backbone.*

The peptide bond has two resonance forms that contribute some double-bond character and inhibit rotation around its axis, so that the alpha carbons are roughly coplanar. The other two dihedral angles in the peptide bond determine the local shape assumed by the protein backbone. The end of the protein with a free carboxyl group is known as the C-terminus or carboxy terminus, whereas the end with a free amino group is known as the N-terminus or amino terminus. The words *protein, polypeptide,* and *peptide* are a little ambiguous and can overlap in meaning. *Protein* is generally used to refer to the complete biological molecule in a stable conformation, whereas *peptide* is generally reserved for a short amino acid oligomers often lacking a stable three-dimensional structure. However, the boundary between the two is not well defined and usually lies near 20–30 residues. *Polypeptide* can refer to any single linear chain of amino acids, usually regardless of length, but often implies an absence of a defined conformation.

Abundance in Cells

It has been estimated that average-sized bacteria contain about 2 million proteins per cell (e.g. *E. coli* and *Staphylococcus aureus*). Smaller bacteria, such as *Mycoplasma* or *spirochetes* contain fewer molecules, namely on the order of 50,000 to 1 million. By contrast, eukaryotic cells are larger and thus contain much more protein. For instance, yeast cells were estimated to contain about 50 million proteins and human cells on the order of 1 to 3 billion. Bacterial genomes encode about 10 times fewer proteins than humans (e.g. small bacteria ~1,000, E. coli: ~4,000, yeast: ~6,000, human: ~20,000).

The concentration of individual proteins ranges from a few molecules per cell to hundreds of thousands, and about a third of all proteins is not produced in most cells or only induced under certain circumstances. For instance, of the 20,000 or so proteins encoded by the human genome only 6,000 are detected in lymphoblastoid cells.

Synthesis

Biosynthesis

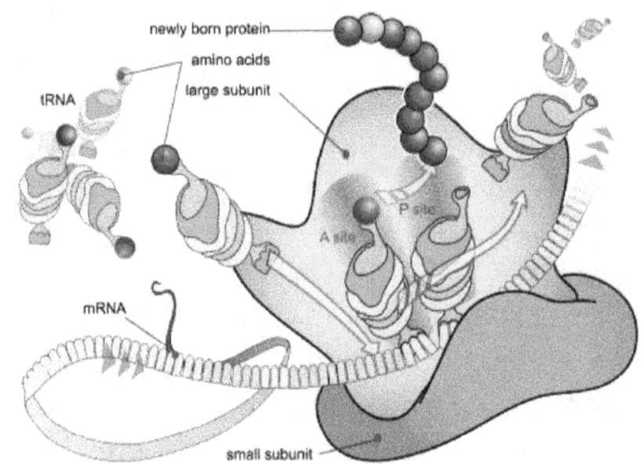

A ribosome produces a protein using mRNA as template

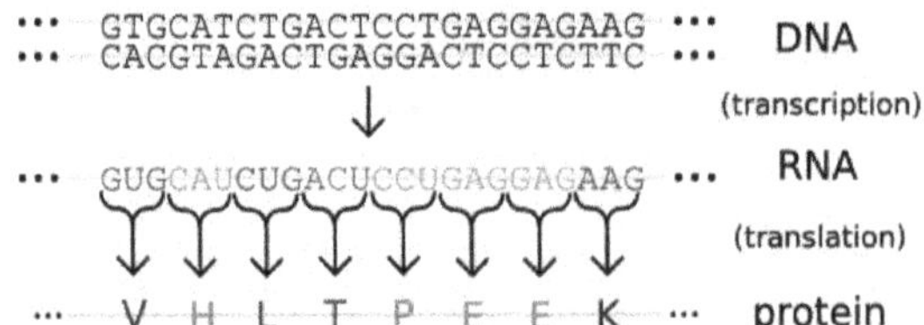

The DNA sequence of a gene encodes the amino acid sequence of a protein

Proteins are assembled from amino acids using information encoded in genes. Each protein has its own unique amino acid sequence that is specified by the nucleotide sequence of the gene encoding this protein. The genetic code is a set of three-nucleotide sets called codons and each three-nucleotide combination designates an amino acid, for example AUG (adenine-uracil-guanine) is the code for methionine. Because DNA contains four nucleotides, the total number of possible codons is 64; hence, there is some redundancy in the genetic code, with some amino acids specified by more than one codon. Genes encoded in DNA are first transcribed into pre-messenger RNA (mRNA) by proteins such as RNA polymerase. Most organisms then process the pre-mRNA (also known as a *primary transcript*) using various forms of Post-transcriptional modification to form the mature mRNA, which is then used as a template for protein synthesis by the ribosome. In prokaryotes the mRNA may either be used as soon as it is produced, or be bound by a ribosome after having moved away from the nucleoid. In contrast, eukaryotes make mRNA in the cell nucleus and then translocate it across the nuclear membrane into the cytoplasm, where protein synthesis then takes place. The rate of protein synthesis is higher in prokaryotes than eukaryotes and can reach up to 20 amino acids per second.

The process of synthesizing a protein from an mRNA template is known as translation. The mRNA is loaded onto the ribosome and is read three nucleotides at a time by matching each codon to its base pairing anticodon located on a transfer RNA molecule, which carries the amino acid corresponding to the codon it recognizes. The enzyme aminoacyl tRNA synthetase "charges" the tRNA molecules with the correct amino acids. The growing polypeptide is often termed the *nascent chain*. Proteins are always biosynthesized from N-terminus to C-terminus.

The size of a synthesized protein can be measured by the number of amino acids it contains and by its total molecular mass, which is normally reported in units of *daltons* (synonymous with atomic mass units), or the derivative unit kilodalton (kDa). Yeast proteins are on average 466 amino acids long and 53 kDa in mass. The largest known proteins are the titins, a component of the muscle sarcomere, with a molecular mass of almost 3,000 kDa and a total length of almost 27,000 amino acids.

Chemical Synthesis

Short proteins can also be synthesized chemically by a family of methods known as peptide synthesis, which rely on organic synthesis techniques such as chemical ligation to produce peptides in high yield. Chemical synthesis allows for the introduction of non-natural amino acids into polypeptide chains, such as attachment of fluorescent probes to amino acid side chains. These methods are useful in laboratory biochemistry and cell biology, though generally not for commercial applications. Chemical synthesis is inefficient for polypeptides longer than about 300 amino acids, and the synthesized proteins may not readily assume their native tertiary structure. Most chemical synthesis methods proceed from C-terminus to N-terminus, opposite the biological reaction.

Structure

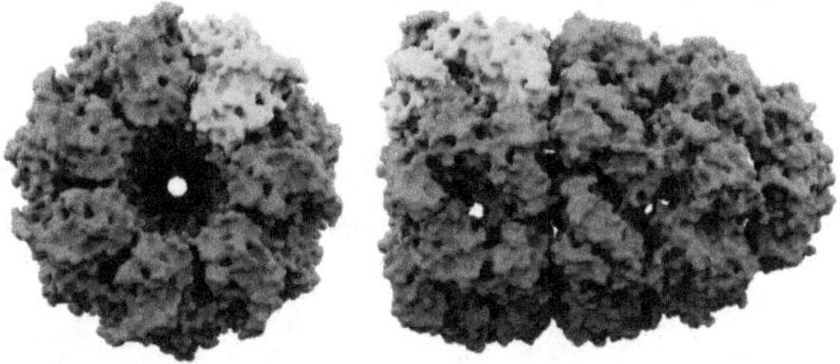

The crystal structure of the chaperonin, a huge protein complex. A single protein subunit is highlighted. Chaperonins assist protein folding.

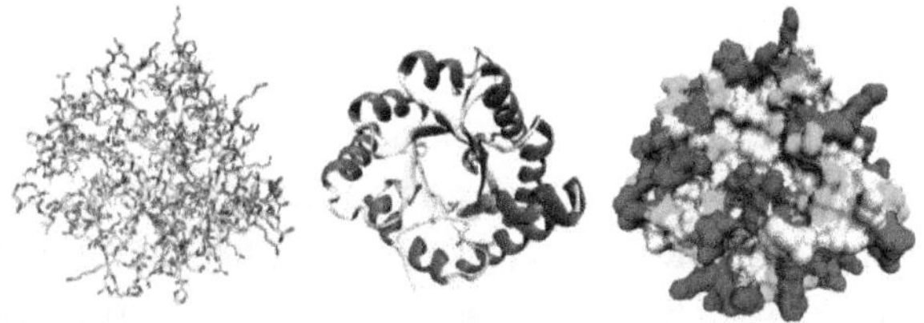

Three possible representations of the three-dimensional structure of the protein triose phosphate isomerase. Left: All-atom representation colored by atom type. Middle: Simplified representation illustrating the backbone conformation, colored by secondary structure. Right: Solvent-accessible surface representation colored by residue type (acidic residues red, basic residues blue, polar residues green, nonpolar residues white).

Most proteins fold into unique 3-dimensional structures. The shape into which a protein naturally folds is known as its native conformation. Although many proteins can fold unassisted, simply through the chemical properties of their amino acids, others require the aid of molecular chaperones to fold into their native states. Biochemists often refer to four distinct aspects of a protein's structure:

- *Primary structure*: the amino acid sequence. A protein is a polyamide.

- *Secondary structure*: regularly repeating local structures stabilized by hydrogen bonds. The most common examples are the α-helix, β-sheet and turns. Because secondary structures are local, many regions of different secondary structure can be present in the same protein molecule.
- *Tertiary structure*: the overall shape of a single protein molecule; the spatial relationship of the secondary structures to one another. Tertiary structure is generally stabilized by non-local interactions, most commonly the formation of a hydrophobic core, but also through salt bridges, hydrogen bonds, disulfide bonds, and even posttranslational modifications. The term "tertiary structure" is often used as synonymous with the term *fold*. The tertiary structure is what controls the basic function of the protein.
- *Quaternary structure*: the structure formed by several protein molecules (polypeptide chains), usually called *protein subunits* in this context, which function as a single protein complex.

Proteins are not entirely rigid molecules. In addition to these levels of structure, proteins may shift between several related structures while they perform their functions. In the context of these functional rearrangements, these tertiary or quaternary structures are usually referred to as "conformations", and transitions between them are called *conformational changes*. Such changes are often induced by the binding of a substrate molecule to an enzyme's active site, or the physical region of the protein that participates in chemical catalysis. In solution proteins also undergo variation in structure through thermal vibration and the collision with other molecules.

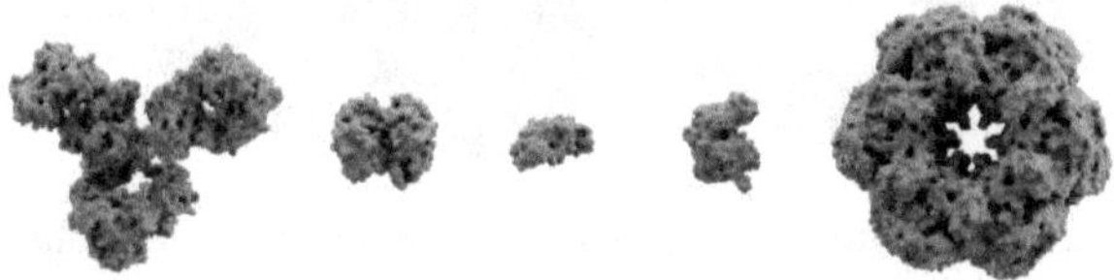

Molecular surface of several proteins showing their comparative sizes. From left to right are: immunoglobulin G (IgG, an antibody), hemoglobin, insulin (a hormone), adenylate kinase (an enzyme), andglutamine synthetase (an enzyme).

Proteins can be informally divided into three main classes, which correlate with typical tertiary structures: globular proteins, fibrous proteins, and membrane proteins. Almost all globular proteins are soluble and many are enzymes. Fibrous proteins are often structural, such as collagen, the major component of connective tissue, or keratin, the protein component of hair and nails. Membrane proteins often serve as receptors or provide channels for polar or charged molecules to pass through the cell membrane.

A special case of intramolecular hydrogen bonds within proteins, poorly shielded from water attack and hence promoting their own dehydration, are called dehydrons.

Structure Determination

Discovering the tertiary structure of a protein, or the quaternary structure of its complexes, can provide important clues about how the protein performs its function. Common experimental methods of structure determination include X-ray crystallography and NMR spectroscopy, both of which can produce information at atomic resolution. However, NMR experiments are able to provide information from which a subset of distances between pairs of atoms can be estimated, and the final possible conformations for a protein are determined by solving a distance geometry problem. Dual polarisation interferometry is a quantitative analytical method for measuring the overall protein conformation and conformational changes due to interactions or other stimulus. Circular dichroism is another laboratory technique for determining internal β-sheet / α-helical composition of proteins. Cryoelectron microscopy is used to produce lower-resolution structural information about very large protein complexes, including assembled viruses; a variant known as electron crystallography can also produce high-resolution information in some cases, especially for two-dimensional crystals of membrane proteins. Solved structures are usually deposited in the Protein Data Bank (PDB), a freely available resource from which structural data about thousands of proteins can be obtained in the form of Cartesian coordinates for each atom in the protein.

Many more gene sequences are known than protein structures. Further, the set of solved structures is biased toward proteins that can be easily subjected to the conditions required in X-ray crystallography, one of the major structure determination methods. In particular, globular proteins are comparatively easy to crystallize in preparation for X-ray crystallography. Membrane proteins, by contrast, are difficult to crystallize and are underrepresented in the PDB. Structural genomics initiatives have attempted to remedy these deficiencies by systematically solving representative structures of major fold classes. Protein structure prediction methods attempt to provide a means of generating a plausible structure for proteins whose structures have not been experimentally determined.

Cellular Functions

Proteins are the chief actors within the cell, said to be carrying out the duties specified by the information encoded in genes. With the exception of certain types of RNA, most other biological molecules are relatively inert elements upon which proteins act. Proteins make up half the dry weight of an *Escherichia coli* cell, whereas other macromolecules such as DNA and RNA make up only 3% and 20%, respectively. The set of proteins expressed in a particular cell or cell type is known as its proteome.

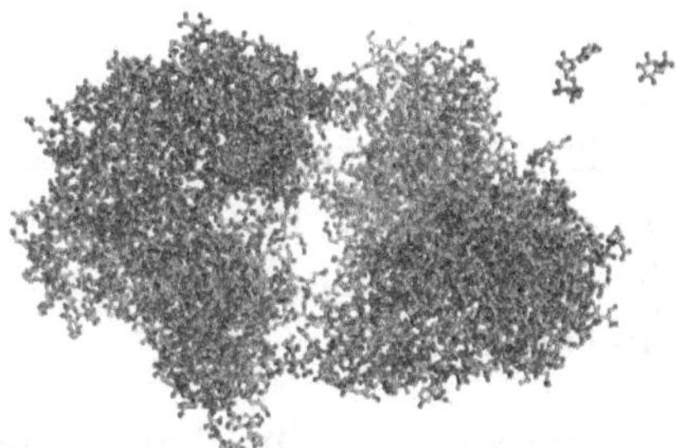

The enzyme hexokinase is shown as a conventional ball-and-stick molecular model. To scale in the top right-hand corner are two of its substrates, ATP and glucose.

The chief characteristic of proteins that also allows their diverse set of functions is their ability to bind other molecules specifically and tightly. The region of the protein responsible for binding another molecule is known as the binding site and is often a depression or "pocket" on the molecular surface. This binding ability is mediated by the tertiary structure of the protein, which defines the binding site pocket, and by the chemical properties of the surrounding amino acids' side chains.

Protein binding can be extraordinarily tight and specific; for example, the ribonuclease inhibitor protein binds to human angiogenin with a sub-femtomolar dissociation constant (<10^{-15} M) but does not bind at all to its amphibian homolog onconase (>1 M). Extremely minor chemical changes such as the addition of a single methyl group to a binding partner can sometimes suffice to nearly eliminate binding; for example, the aminoacyl tRNA synthetase specific to the amino acid valine discriminates against the very similar side chain of the amino acid isoleucine.

Proteins can bind to other proteins as well as to small-molecule substrates. When proteins bind specifically to other copies of the same molecule, they can oligomerize to form fibrils; this process occurs often in structural proteins that consist of globular monomers that self-associate to form rigid fibers. Protein–protein interactions also regulate enzymatic activity, control progression through the cell cycle, and allow the assembly of large protein complexes that carry out many closely related reactions with a common biological function. Proteins can also bind to, or even be integrated into, cell membranes. The ability of binding partners to induce conformational changes in proteins allows the construction of enormously complex signaling networks. As interactions between proteins are reversible, and depend heavily on the availability of different groups of partner proteins to form aggregates that are capable to carry out discrete sets of function, study of the interactions between specific proteins is a key to understand important aspects of cellular function, and ultimately the properties that distinguish particular cell types.

Enzymes

The best-known role of proteins in the cell is as enzymes, which catalyse chemical reactions. Enzymes are usually highly specific and accelerate only one or a few chemical reactions. Enzymes carry out most of the reactions involved in metabolism, as well as manipulating DNA in processes such as DNA replication, DNA repair, and transcription. Some enzymes act on other proteins to add or remove chemical groups in a process known as posttranslational modification. About 4,000 reactions are known to be catalysed by enzymes. The rate acceleration conferred by enzymatic catalysis is often enormous—as much as 10^{17}-fold increase in rate over the uncatalysed reaction in the case of orotate decarboxylase (78 million years without the enzyme, 18 milliseconds with the enzyme).

The molecules bound and acted upon by enzymes are called substrates. Although enzymes can consist of hundreds of amino acids, it is usually only a small fraction of the residues that come in contact with the substrate, and an even smaller fraction—three to four residues on average—that are directly involved in catalysis. The region of the enzyme that binds the substrate and contains the catalytic residues is known as the active site.

Dirigent proteins are members of a class of proteins that dictate the stereochemistry of a compound synthesized by other enzymes.

Cell Signaling and Ligand Binding

Many proteins are involved in the process of cell signaling and signal transduction. Some proteins, such as insulin, are extracellular proteins that transmit a signal from the cell in which they were synthesized to other cells in distant tissues. Others are membrane proteins that act as receptors whose main function is to bind a signaling molecule and induce a biochemical response in the cell.

Many receptors have a binding site exposed on the cell surface and an effector domain within the cell, which may have enzymatic activity or may undergo a conformational change detected by other proteins within the cell.

Ribbon diagram of a mouse antibody against cholera that binds a carbohydrate antigen

Antibodies are protein components of an adaptive immune system whose main function is to bind antigens, or foreign substances in the body, and target them for destruction. Antibodies can be secreted into the extracellular environment or anchored in the membranes of specialized B cells known as plasma cells. Whereas enzymes are limited in their binding affinity for their substrates by the necessity of conducting their reaction, antibodies have no such constraints. An antibody's binding affinity to its target is extraordinarily high.

Many ligand transport proteins bind particular small biomolecules and transport them to other locations in the body of a multicellular organism. These proteins must have a high binding affinity when their ligand is present in high concentrations, but must also release the ligand when it is present at low concentrations in the target tissues. The canonical example of a ligand-binding protein is haemoglobin, which transports oxygen from the lungs to other organs and tissues in all vertebrates and has close homologs in every biological kingdom. Lectins are sugar-binding proteins which are highly specific for their sugar moieties. Lectins typically play a role in biological recognition phenomena involving cells and proteins. Receptors and hormones are highly specific binding proteins.

Transmembrane proteins can also serve as ligand transport proteins that alter the permeability of the cell membrane to small molecules and ions. The membrane alone has a hydrophobic core through which polar or charged molecules cannot diffuse. Membrane proteins contain internal channels that allow such molecules to enter and exit the cell. Many ion channel proteins are specialized to select for only a particular ion; for example, potassium and sodium channels often discriminate for only one of the two ions.

Structural Proteins

Structural proteins confer stiffness and rigidity to otherwise-fluid biological components. Most structural proteins are fibrous proteins; for example, collagen and elastin are critical components of connective tissue such as cartilage, and keratin is found in hard or filamentous structures such as hair, nails, feathers, hooves, and some animal shells. Some globular proteins can also play structural functions, for example, actin and tubulin are globular and soluble as monomers, but polymerize to form long, stiff fibers that make up the cytoskeleton, which allows the cell to maintain its shape and size.

Other proteins that serve structural functions are motor proteins such as myosin, kinesin, and dynein, which are capable of generating mechanical forces. These proteins are crucial for cellular motility of single celled organisms and the sperm of many multicellular organisms which reproduce sexually. They also generate the forces exerted by contracting muscles and play essential roles in intracellular transport.

Methods of Study

The activities and structures of proteins may be examined *in vitro, in vivo, and in silico. In vitro* studies of purified proteins in controlled environments are useful for learning how a protein carries out its function: for example, enzyme kinetics studies explore the chemical mechanism of an enzyme's catalytic activity and its relative affinity for various possible substrate molecules. By contrast, *in vivo* experiments can provide information about the physiological role of a protein in the context of a cell or even a whole organism. *In silico* studies use computational methods to study proteins.

Protein Purification

To perform *in vitro* analysis, a protein must be purified away from other cellular components. This process usually begins with cell lysis, in which a cell's membrane is disrupted and its internal contents released into a solution known as a crude lysate. The resulting mixture can be purified using ultracentrifugation, which fractionates the various cellular components into fractions containing soluble proteins; membrane lipids and proteins; cellular organelles, and nucleic acids. Precipitation by a method known as salting out can concentrate the proteins from this lysate. Various types of chromatography are then used to isolate the protein or proteins of interest based on properties such as molecular weight, net charge and binding affinity. The level of purification can be monitored using various types of gel electrophoresis if the desired protein's molecular weight and isoelectric point are known, by spectroscopy if the protein has distinguishable spectroscopic features, or by enzyme assays if the protein has enzymatic activity. Additionally, proteins can be isolated according their charge using electrofocusing.

For natural proteins, a series of purification steps may be necessary to obtain protein sufficiently pure for laboratory applications. To simplify this process, genetic engineering is often used to add chemical features to proteins that make them easier to purify without affecting their structure or activity. Here, a "tag" consisting of a specific amino acid sequence, often a series of histidine residues (a "His-tag"), is attached to one terminus of the protein. As a result, when the lysate is passed over a chromatography column containing nickel, the histidine residues ligate the nickel and attach to the column while the untagged components of the lysate pass unimpeded. A number of different tags have been developed to help researchers purify specific proteins from complex mixtures.

Cellular Localization

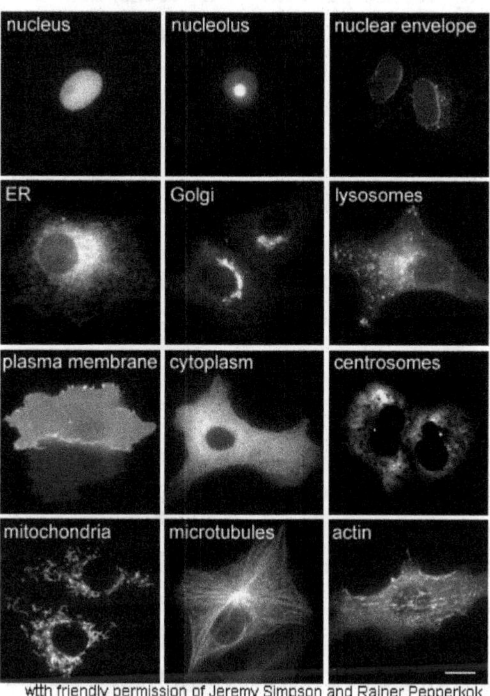

Proteins in different cellular compartments and structures tagged with green fluorescent protein (here, white)

The study of proteins *in vivo* is often concerned with the synthesis and localization of the protein within the cell. Although many intracellular proteins are synthesized in the cytoplasm and membrane-bound or secreted proteins in the endoplasmic reticulum, the specifics of how proteins are targeted to specific organelles or cellular structures is often unclear. A useful technique for assessing cellular localization uses genetic engineering to express in a cell a fusion protein or chimera consisting of the natural protein of interest linked to a "reporter" such as green fluorescent protein (GFP). The fused protein's position within the cell can be cleanly and efficiently visualized using microscopy, as shown in the figure opposite.

Other methods for elucidating the cellular location of proteins requires the use of known compartmental markers for regions such as the ER, the Golgi, lysosomes or vacuoles, mitochondria, chloroplasts, plasma membrane, etc. With the use of fluorescently tagged versions of these markers or of antibodies to known markers, it becomes much simpler to identify the localization of a protein of interest. For example, indirect immunofluorescence will allow for fluorescence colocalization and demonstration of location. Fluorescent dyes are used to label cellular compartments for a similar purpose.

Other possibilities exist, as well. For example, immunohistochemistry usually utilizes an antibody to one or more proteins of interest that are conjugated to enzymes yielding either luminescent or chromogenic signals that can be compared between samples, allowing for localization information. Another applicable technique is cofractionation in sucrose (or other material) gradients using isopycnic centrifugation. While this technique does not prove colocalization of a compartment of known density and the protein of interest, it does increase the likelihood, and is more amenable to large-scale studies.

Finally, the gold-standard method of cellular localization is immunoelectron microscopy. This technique also uses an antibody to the protein of interest, along with classical electron microscopy techniques. The sample is prepared for normal electron microscopic examination, and then treated with an antibody to the protein of interest that is conjugated to an extremely electro-dense material, usually gold. This allows for the localization of both ultrastructural details as well as the protein of interest.

Through another genetic engineering application known as site-directed mutagenesis, researchers can alter the protein sequence and hence its structure, cellular localization, and susceptibility to regulation. This technique even allows the incorporation of unnatural amino acids into proteins, using modified tRNAs, and may allow the rational design of new proteins with novel properties.

Proteomics

The total complement of proteins present at a time in a cell or cell type is known as its proteome, and the study of such large-scale data sets defines the field of proteomics, named by analogy to the related field of genomics. Key experimental techniques in proteomics include 2D electrophoresis, which allows the separation of a large number of proteins, mass spectrometry, which allows rapid high-throughput identification of proteins and sequencing of peptides (most often after in-gel digestion), protein microarrays, which allow the detection of the relative levels of a large number of proteins present in a cell, and two-hybrid screening, which allows the systematic exploration of protein–protein interactions. The total complement of biologically possible such interactions is known as the interactome. A systematic attempt to determine the structures of proteins representing every possible fold is known as structural genomics.

Bioinformatics

A vast array of computational methods have been developed to analyze the structure, function, and evolution of proteins.

The development of such tools has been driven by the large amount of genomic and proteomic data available for a variety of organisms, including the human genome. It is simply impossible to study all proteins experimentally, hence only a few are subjected to laboratory experiments while computational tools are used to extrapolate to similar proteins. Such homologous proteins can be efficiently identified in distantly related organisms by sequence alignment. Genome and gene sequences can be searched by a variety of tools for certain properties. Sequence profiling tools can find restriction enzyme sites, open reading frames in nucleotide sequences, and predict secondary structures. Phylogenetic trees can be constructed and evolutionary hypotheses developed using special software like ClustalW regarding the ancestry of modern organisms and the genes they express. The field of bioinformatics is now indispensable for the analysis of genes and proteins.

Structure Prediction and Simulation

Complementary to the field of structural genomics, protein structure prediction seeks to develop efficient ways to provide plausible models for proteins whose structures have not yet been determined experimentally. The most successful type of structure prediction, known as homology modeling, relies on the existence of a "template" structure with sequence similarity to the protein being modeled; structural genomics' goal is to provide sufficient representation in solved structures to model most of those that remain. Although producing accurate models remains a challenge when only distantly related template structures are available, it has been suggested that sequence alignment is the bottleneck in this process, as quite accurate models can be produced if a "perfect" sequence alignment is known. Many structure prediction methods have served to inform the emerging field of protein engineering, in which novel protein folds have already been designed. A more complex computational problem is the prediction of intermolecular interactions, such as in molecular docking and protein–protein interaction prediction.

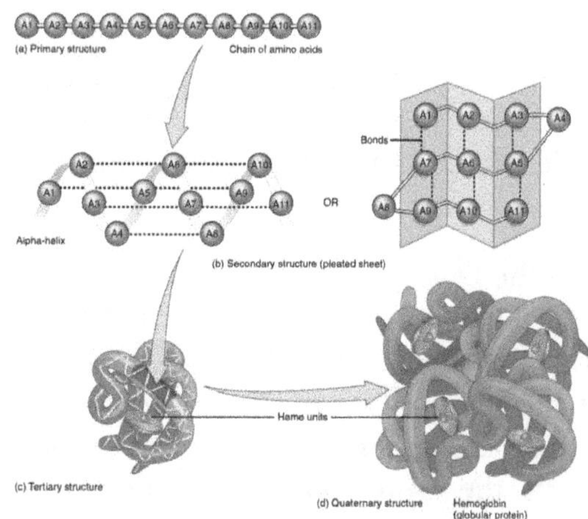

Constituent amino-acids can be analyzed to predict secondary, tertiary and quaternary protein structure, in this case hemoglobin containing heme units

The processes of protein folding and binding can be simulated using such technique as molecular mechanics, in particular, molecular dynamics and Monte Carlo, which increasingly take advantage of parallel and distributed computing (Folding@home project; molecular modeling on GPU). The folding of small α-helical protein domains such as the villin headpiece and the HIV accessory protein have been successfully simulated *in silico*, and hybrid methods that combine standard molecular dynamics with quantum mechanics calculations have allowed exploration of the electronic states of rhodopsins.

Protein Disorder and Unstructure Prediction

Many proteins (in Eucaryota ~33%) contain large unstructured but biologically functional segments and can be classified as intrinsically disordered proteins. Predicting and analysing protein disorder is, therefore, an important part of protein structure characterisation.

Nutrition

Most microorganisms and plants can biosynthesize all 20 standard amino acids, while animals (including humans) must obtain some of the amino acids from the diet. The amino acids that an organism cannot synthesize on its own are referred to as essential amino acids. Key enzymes that synthesize certain amino acids are not present in animals — such as aspartokinase, which catalyses the first step in the synthesis of lysine, methionine, and threonine from aspartate. If amino acids are present in the environment, microorganisms can conserve energy by taking up the amino acids from their surroundings and downregulating their biosynthetic pathways.

In animals, amino acids are obtained through the consumption of foods containing protein. Ingested proteins are then broken down into amino acids through digestion, which typically involves denaturation of the protein through exposure to acid and hydrolysis by enzymes called proteases. Some ingested amino acids are used for protein biosynthesis, while others are converted to glucose through gluconeogenesis, or fed into the citric acid cycle. This use of protein as a fuel is particularly important under starvation conditions as it allows the body's own proteins to be used to support life, particularly those found in muscle. Amino acids are also an important dietary source of nitrogen.

History and Etymology

Proteins were recognized as a distinct class of biological molecules in the eighteenth century by Antoine Fourcroy and others, distinguished by the molecules' ability to coagulate or flocculate under treatments with heat or acid. Noted examples at the time included albumin from egg whites, blood serum albumin, fibrin, and wheat gluten.

Proteins were first described by the Dutch chemist Gerardus Johannes Mulder and named by the Swedish chemist Jöns Jacob Berzelius in 1838. Mulder carried out elemental analysis of common proteins and found that nearly all proteins had the same empirical formula, $C_{400}H_{620}N_{100}O_{120}P_1S_1$. He came to the erroneous conclusion that they might be composed of a single type of (very large) molecule. The term "protein" to describe these molecules was proposed by Mulder's associate Berzelius; protein is derived from the word (*proteios*), meaning "primary", "in the lead", or "standing in front", + *-in*. Mulder went on to identify the products of protein degradation such as the amino acid leucine for which he found a (nearly correct) molecular weight of 131 Da.

Early nutritional scientists such as the German Carl von Voit believed that protein was the most important nutrient for maintaining the structure of the body, because it was generally believed that "flesh makes flesh." Karl Heinrich Ritthausen extended known protein forms with the identification of glutamic acid. At the Connecticut Agricultural Experiment Station a detailed review of the vegetable proteins was compiled by Thomas Burr Osborne. Working with Lafayette Mendel and applying Liebig's law of the minimum in feeding laboratory rats, the nutritionally essential amino acids were established. The work was continued and communicated by William Cumming Rose. The understanding of proteins as polypeptides came through the work of Franz Hofmeister and Hermann Emil Fischer. The central role of proteins as enzymes in living organisms was not fully appreciated until 1926, when James B. Sumner showed that the enzyme urease was in fact a protein.

The difficulty in purifying proteins in large quantities made them very difficult for early protein biochemists to study. Hence, early studies focused on proteins that could be purified in large quantities, e.g., those of blood, egg white, various toxins, and digestive/metabolic enzymes obtained from slaughterhouses. In the 1950s, the Armour Hot Dog Co. purified 1 kg of pure bovine pancreatic ribonuclease A and made it freely available to scientists; this gesture helped ribonuclease A become a major target for biochemical study for the following decades.

Linus Pauling is credited with the successful prediction of regular protein secondary structures based on hydrogen bonding, an idea first put forth by William Astbury in 1933. Later work by Walter Kauzmann on denaturation, based partly on previous studies by Kaj Linderstrøm-Lang, contributed an understanding of protein folding and structure mediated by hydrophobic interactions.

John Kendrew with model of myoglobin in progress

The first protein to be sequenced was insulin, by Frederick Sanger, in 1949. Sanger correctly determined the amino acid sequence of insulin, thus conclusively demonstrating that proteins consisted of linear polymers of amino acids rather than branched chains, colloids, or cyclols. He won the Nobel Prize for this achievement in 1958.

The first protein structures to be solved were hemoglobin and myoglobin, by Max Perutz and Sir John Cowdery Kendrew, respectively, in 1958. As of 2016, the Protein Data Bank has over 115,000 atomic-resolution structures of proteins. In more recent times, cryo-electron microscopy of large macromolecular assemblies and computational protein structure prediction of small protein domains are two methods approaching atomic resolution.

Techniques used to Study Protein

Protein Methods

Protein methods are the techniques used to study proteins. There are experimental methods for studying proteins (e.g., for detecting proteins, for isolating and purifying proteins, and for characterizing the structure and function of proteins, often requiring that the protein first be purified). Computational methods typically use computer programs to analyze proteins. However, many experimental methods (e.g., mass spectrometry) require computational analysis of the raw data.

Genetic Methods

Experimental analysis of proteins typically requires expression and purification of proteins. Expression is achieved by manipulating DNA that encodes the protein(s) of interest. Hence, protein analysis usually requires DNA methods, especially cloning. Some examples of genetic methods include conceptual translation, Site-directed mutagenesis, using a fusion protein, and matching allele with disease states. Some proteins have never been directly sequenced, however by translating codons from known mRNA sequences into amino acids by a method known as conceptual translation. Site-directed mutagenesis selectively introduces mutations that change the structure of a protein. The function of parts of proteins can be better understood by studying the change in phenotype as a result of this change. Fusion proteins are made by inserting protein tags, such as the His-tag, to produce a modified protein that is easier to track. An example of this would be GFP-Snf2H which consists of a protein bound to a green fluorescent protein to form a hybrid protein. By analyzing DNA alleles can be identified as being associated with disease states, such as in calculation of LOD scores.

Protein Extraction from Tissues

Protein extraction from tissues with tough extracellular matrices (e.g., biopsy samples, venous tissues, cartilage, skin) is often achieved in a laboratory setting by impact pulverization in liquid nitrogen. Samples are frozen in liquid nitrogen and subsequently subjected to impact or mechanical grinding. As water in the samples becomes very brittle at these temperature, the samples are often reduced to a collection of fine fragments, which can then be dissolved for protein extraction. Stainless steel devices known as tissue pulverizers are sometimes used for this purpose.

Advantages of these devices include high levels of protein extraction from small, valuable samples, disadvantages include low-level cross-over contamination.

Protein Purification

- Protein isolation
    - Chromatography methods: ion exchange, size-exclusion chromatography (or gel filtration), affinity chromatography
- Protein extraction and solubilization
- Concentrating protein solutions
- Gel electrophoresis
    - Gel electrophoresis under denaturing conditions
    - Gel electrophoresis under non-denaturing conditions
    - 2D gel electrophoresis
- Electrofocusing

Detecting Proteins

Non-specific Methods that Detect Total Protein Only

- Absorbance: Read at 280 or 205 nm. Can be very inaccurate. Detection in the range of 100 µg/mL to 1 mg/mL
- Bradford protein assay: Detection in the range of ~1 mg/mL
- Biuret Test Derived Assays:
    - Bicinchoninic acid assay (BCA assay): Detection down to 0.5 µg/mL
    - Lowry Protein assay: Detection in the range of 0.01–1.0 mg/mL
- Fluorescamine: Quantifies proteins and peptides in solution if primary amine are present in the amino acids
- Amido black: Detection in the range of 1-12 µg/mL

- Colloidal gold: Detection in the range of 20 - 640 ng/mL
- Nitrogen detection:
    - Kjeldahl method: used primarily for food and requires oxidation of material
    - Dumas method: used primarily for food and requires combustion of material

Specific Methods Which Can Detect Amount of a Single Protein

- Spectrometry methods:
    - High-performance liquid chromatography (HPLC): Chromatography method to detect proteins or peptides
    - Liquid chromatography–mass spectrometry (LC/MS): Can detect proteins at low concentrations (ng/mL to pg/mL) in blood and body fluids, such as for Pharmacokinetics.
- Antibody dependent methods:
    - Enzyme-linked immunosorbent assay (ELISA): Specifically can detect protein down to pg/mL.
    - Protein immunoprecipitation: technique of precipitating a protein antigen out of solution using an antibody that specifically binds to that particular protein.
    - Immunoelectrophoresis: separation and characterization of proteins based on electrophoresis and reaction with antibodies.
    - Western blot: couples gel electrophoresis and incubation with antibodies to detect specific proteins in a sample of tissue homogenate or extract (a type of Immunoelectrophoresis technique).
    - Protein immunostaining

Protein Structures

- X-ray crystallography
- Protein NMR
- Cryo-electron microscopy
- Small-angle X-ray scattering

Interactions Involving Proteins

- Protein footprinting

Protein–protein Interactions

- (Yeast) two-hybrid system

- Protein-fragment complementation assay
- Co-immunoprecipitation
- Affinity purification and mass spectrometry

Protein–DNA Interactions

- ChIP-on-chip
- Chip-sequencing
- DamID
- Microscale thermophoresis

Protein–RNA Interactions

- Toeprinting assay
- TCP-seq

Computational Methods

- Molecular dynamics
- Protein structure prediction
- Protein sequence alignment (sequence comparison, including BLAST)
- Protein structural alignment
- Protein ontology

Other Methods

- Hydrogen–deuterium exchange
- Mass spectrometry
- Protein sequencing
- Protein synthesis
- Proteomics
- Peptide mass fingerprinting
- Ligand binding assay
- Eastern blotting
- Metabolic labeling

- Heavy isotope labeling
- Radioactive isotope labeling

Protein Sequencing

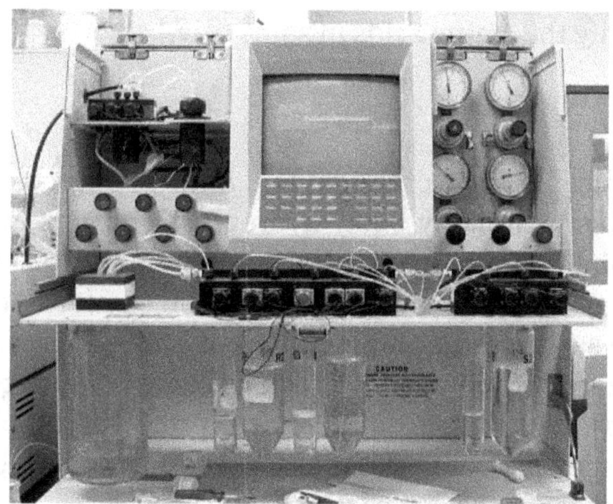

A Beckman-Coulter Porton LF3000G protein sequencing machine

Protein sequencing is a technique to determine the amino acid sequence of a protein, as well as which conformation the protein adopts and the extent to which it is complexed with any non-peptide molecules. Discovering the structures and functions of proteins in living organisms is an important tool for understanding cellular processes, and allows drugs that target specific metabolic pathways to be invented more easily.

The two major direct methods of protein sequencing are mass spectrometry and the Edman degradation reaction. It is also possible to generate an amino acid sequence from the DNA or mRNA sequence encoding the protein, if this is known. However, there are many other reactions that can be used to gain more limited information about protein sequences and can be used as preliminaries to the aforementioned methods of sequencing or to overcome specific inadequacies within them.

Protein Sequencer

A protein sequencer is a machine that is used to determine the sequence of amino acids in a protein.

They work by tagging and removing one amino acid at a time, which is analysed and identified. This is done repetitively for the whole polypeptide, until the whole sequence is established.

This method has generally been replaced by nucleic acid technology, and it is often easier to identify the sequence of a protein by looking at the DNA that codes for it.

Determining Amino Acid Composition

It is often desirable to know the unordered amino acid composition of a protein prior to attempting to find the ordered sequence, as this knowledge can be used to facilitate the discovery of errors in the sequencing process or to distinguish between ambiguous results. Knowledge of the frequency of certain amino acids may also be used to choose which protease to use for digestion of the protein. A generalized method often referred to as *amino acid analysis* for determining amino acid frequency is as follows:

1. Hydrolyse a known quantity of protein into its constituent amino acids.
2. Separate and quantify the amino acids in some way.

Hydrolysis

Hydrolysis is done by heating a sample of the protein in 6 M hydrochloric acid to 100–110 °C for 24 hours or longer. Proteins with many bulky hydrophobic groups may require longer heating periods. However, these conditions are so vigorous that some amino acids (serine, threonine, tyrosine, tryptophan, glutamine, and cysteine) are degraded. To circumvent this problem, Biochemistry Online suggests heating separate samples for different times, analysing each resulting solution, and extrapolating back to zero hydrolysis time. Rastall suggests a variety of reagents to prevent or reduce degradation, such as thiol reagents or phenol to protect tryptophan and tyrosine from attack by chlorine, and pre-oxidising cysteine. He also suggests measuring the quantity of ammonia evolved to determine the extent of amide hydrolysis.

Separation

The amino acids can be separated by ion-exchange chromatography or hydrophobic interaction chromatography. An example of the former is given by the NTRC using sulfonated polystyrene as a matrix, adding the amino acids in acid solution and passing a buffer of steadily increasing pH through the column. Amino acids will be eluted when the pH reaches their respective isoelectric points. The latter technique may be employed through the use of reversed phase chromatography. Many commercially available C8 and C18 silica columns have demonstrated successful separation of amino acids in solution in less than 40 minutes through the use of an optimised elution gradient.

Quantitative Analysis

Once the amino acids have been separated, their respective quantities are determined by adding a reagent that will form a coloured derivative. If the amounts of amino acids are in excess of 10 nmol, ninhydrin can be used for this; it gives a yellow colour when reacted with proline, and a vivid purple with other amino acids. The concentration of amino acid is proportional to the absorbance of the resulting solution. With very small quantities, down to 10 pmol, fluorescamine can be used as a marker: This forms a fluorescent derivative on reacting with an amino acid.

N-terminal Amino Acid Analysis

Determining which amino acid forms the *N*-terminus of a peptide chain is useful for two reasons: to aid the ordering of individual peptide fragments' sequences into a whole chain, and because the first round of Edman degradation is often contaminated by impurities and therefore does not give an accurate determination of the *N*-terminal amino acid. A generalised method for *N*-terminal amino acid analysis follows:

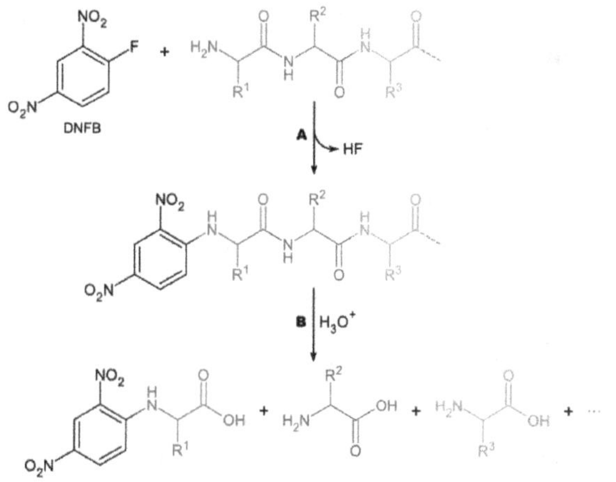

Sanger's method of peptide end-group analysis: **A** derivatization of *N*-terminal end with Sanger's reagent (DNFB), **B** total acid hydrolysis of the dinitrophenyl peptide

1. React the peptide with a reagent that will selectively label the terminal amino acid.
2. Hydrolyse the protein.
3. Determine the amino acid by chromatography and comparison with standards.

There are many different reagents which can be used to label terminal amino acids. They all react with amine groups and will therefore also bind to amine groups in the side chains of amino acids such as lysine - for this reason it is necessary to be careful in interpreting chromatograms to ensure that the right spot is chosen. Two of the more common reagents are Sanger's reagent (1-fluoro-2,4-dinitrobenzene) and dansyl derivatives such as dansyl chloride. Phenylisothiocyanate, the reagent for the Edman degradation, can also be used. The same questions apply here as in the determination of amino acid composition, with the exception that no stain is needed, as the reagents produce coloured derivatives and only qualitative analysis is required. So the amino acid does not have to be eluted from the chromatography column, just compared with a standard. Another consideration to take into account is that, since any amine groups will have reacted with the labelling reagent, ion exchange chromatography cannot be used, and thin layer chromatography or high-pressure liquid chromatography should be used instead.

C-terminal Amino Acid Analysis

The number of methods available for C-terminal amino acid analysis is much smaller than the number of available methods of N-terminal analysis. The most common method is to add carboxypeptidases to a solution of the protein, take samples at regular intervals, and determine the terminal amino acid by analysing a plot of amino acid concentrations against time. This method will be very useful in the case of polypeptides and protein-blocked N termini. C-terminal sequencing would greatly help in verifying the primary structures of proteins predicted from DNA sequences and to detect any postranslational processing of gene products from known codon sequences.

Edman Degradation

The Edman degradation is a very important reaction for protein sequencing, because it allows the ordered amino acid composition of a protein to be discovered. Automated Edman sequencers are now in widespread use, and are able to sequence peptides up to approximately 50 amino acids long. A reaction scheme for sequencing a protein by the Edman degradation follows; some of the steps are elaborated on subsequently.

1. Break any disulfide bridges in the protein with a reducing agent like 2-mercaptoethanol. A protecting group such as iodoacetic acid may be necessary to prevent the bonds from re-forming.
2. Separate and purify the individual chains of the protein complex, if there are more than one.
3. Determine the amino acid composition of each chain.
4. Determine the terminal amino acids of each chain.
5. Break each chain into fragments under 50 amino acids long.
6. Separate and purify the fragments.
7. Determine the sequence of each fragment.
8. Repeat with a different pattern of cleavage.
9. Construct the sequence of the overall protein.

Digestion into Peptide Fragments

Peptides longer than about 50-70 amino acids long cannot be sequenced reliably by the Edman degradation. Because of this, long protein chains need to be broken up into small fragments that can then be sequenced individually. Digestion is done either by endopeptidases such as trypsin or pepsin or by chemical reagents such as cyanogen bromide. Different enzymes give different cleavage patterns, and the overlap between fragments can be used to construct an overall sequence.

The Edman Degradation Reaction

The peptide to be sequenced is adsorbed onto a solid surface. One common substrate is glass fibre coated with polybrene, a cationic polymer. The Edman reagent, phenylisothiocyanate (PITC), is added to the adsorbed peptide, together with a mildly basic buffer solution of 12% trimethylamine. This reacts with the amine group of the N-terminal amino acid.

The terminal amino acid can then be selectively detached by the addition of anhydrous acid. The derivative then isomerises to give a substituted phenylthiohydantoin, which can be washed off and identified by chromatography, and the cycle can be repeated. The efficiency of each step is about 98%, which allows about 50 amino acids to be reliably determined.

Limitations of the Edman Degradation

Because the Edman degradation proceeds from the N-terminus of the protein, it will not work if the N-terminal amino acid has been chemically modified or if it is concealed within the body of the protein. It also requires the use of either guesswork or a separate procedure to determine the positions of disulfide bridges.

Mass Spectrometry

The other major direct method by which the sequence of a protein can be determined is mass spectrometry. This method has been gaining popularity in recent years as new techniques and increasing computing power have facilitated it. Mass spectrometry can, in principle, sequence any size of protein, but the problem becomes computationally more difficult as the size increases. Peptides are also easier to prepare for mass spectrometry than whole proteins, because they are more soluble. One method of delivering the peptides to the spectrometer is electrospray ionization, for which John Bennett Fenn won the Nobel Prize in Chemistry in 2002. The protein is digested by an endoprotease, and the resulting solution is passed through a high-pressure liquid chromatography column. At the end of this column, the solution is sprayed out of a narrow nozzle charged to a high positive potential into the mass spectrometer. The charge on the droplets causes them to fragment until only single ions remain. The peptides are then fragmented and the mass-to-charge ratios of the fragments measured. (It is possible to detect which peaks correspond to multiply charged fragments, because these will have auxiliary peaks corresponding to other isotopes - the distance between these other peaks is inversely proportional to the charge on the fragment). The mass spectrum is analysed by computer and often compared against a database of previously sequenced proteins in order to determine the sequences of the fragments. This process is then repeated with a different digestion enzyme, and the overlaps in the sequences are used to construct a sequence for the protein.

There is a fundamental issue of relying on database search to sequence the protein. The database search algorithms assume the database sequences are correct. Therefore, when comparing the spectra with the database peptides, as long as some evidences are observed, e.g. the precursor mass and the peptide mass are very close and there are some fragment ions matched, it will make the peptide spectrum assignment. But this does not mean that evidences are observed to determine each amino acid.

Predicting Protein Sequence from DNA/RNA Sequences

In organisms that do not have introns (e.g., prokaryotes) the amino acid sequence of a protein can also be determined indirectly from the mRNA or the DNA that codes for the protein. If the sequence of the gene is already known, then this is all very easy. However, it is rare that the DNA sequence of a newly isolated protein will be known, and, so, if this method is to be used, it has to be found in some way. One way that this can be done is to sequence a short section, perhaps 15 amino acids long, of the protein by one of the above methods, and then use this sequence to generate a complementary marker for the protein's RNA. This can then be used to isolate the mRNA coding for the protein, which can then be replicated in a polymerase chain reaction to yield a significant amount of DNA, which can then be sequenced relatively easily. The amino acid sequence of the protein can then be deduced from this. However, it is necessary to take into account the possibility of amino acids being removed after the mRNA has been translated.

Bioinformatics Tools for Sequencing

Bioinformatics tools exist that translate nucleic acid sequences into their corresponding polypeptide chain. Such tool takes an input of a nucleic acid sequence and will produce the corresponding amino acid sequence based on the selected settings.

Another option is to use a bioinformatics tool that takes an amino acid sequence and back-translates it into the most likely nucleic acid sequence.

Mass Spectrometry

SIMS mass spectrometer, model IMS 3f.

////Mass spectrometry (MS) is an analytical technique that ionizes chemical species and sorts the ions based on their mass to charge ratio. In simpler terms, a mass spectrum measures the masses within a sample. Mass spectrometry is used in many different fields and is applied to pure samples as well as complex mixtures.

A mass spectrum is a plot of the ion signal as a function of the mass-to-charge ratio. These spectra are used to determine the elemental or isotopic signature of a sample, the masses of particles and of molecules, and to elucidate the chemical structures of molecules, such as peptides and other chemical compounds.

In a typical MS procedure, a sample, which may be solid, liquid, or gas, is ionized, for example by bombarding it with electrons. This may cause some of the sample's molecules to break into charged fragments. These ions are then separated according to their mass-to-charge ratio, typically by accelerating them and subjecting them to an electric or magnetic field: ions of the same mass-to-charge ratio will undergo the same amount of deflection. The ions are detected by a mechanism capable of detecting charged particles, such as an electron multiplier. Results are displayed as spectra of the relative abundance of detected ions as a function of the mass-to-charge ratio. The atoms or molecules in the sample can be identified by correlating known masses to the identified masses or through a characteristic fragmentation pattern.

History

Replica of J.J. Thompson's third mass spectrometer.

In 1886, Eugen Goldstein observed rays in gas discharges under low pressure that traveled away from the anode and through channels in a perforated cathode, opposite to the direction of negatively charged cathode rays (which travel from cathode to anode). Goldstein called these positively charged anode rays "Kanalstrahlen"; the standard translation of this term into English is "canal rays". Wilhelm Wien found that strong electric or magnetic fields deflected the canal rays and, in 1899, constructed a device with parallel electric and magnetic fields that separated the positive rays according to their charge-to-mass ratio (Q/m). Wien found that the charge-to-mass ratio depended on the nature of the gas in the discharge tube. English scientist J.J. Thomson later improved on the work of Wien by reducing the pressure to create the mass spectrograph.

Calutron mass spectrometers were used in the Manhattan Project for uranium enrichment.

The word *spectrograph* had become part of the international scientific vocabulary by 1884. Early *spectrometry* devices that measured the mass-to-charge ratio of ions were called *mass spectrographs* which consisted of instruments that recorded a spectrum of mass values on a photographic plate. A *mass spectroscope* is similar to a *mass spectrograph* except that the beam of ions is directed onto a phosphor screen. A mass spectroscope configuration was used in early instruments when it was desired that the effects of adjustments be quickly observed. Once the instrument was properly adjusted, a photographic plate was inserted and exposed. The term mass spectroscope continued to be used even though the direct illumination of a phosphor screen was replaced by indirect measurements with an oscilloscope. The use of the term *mass spectroscopy* is now discouraged due to the possibility of confusion with light spectroscopy. Mass spectrometry is often abbreviated as *mass-spec* or simply as *MS*.

Modern techniques of mass spectrometry were devised by Arthur Jeffrey Dempster and F.W. Aston in 1918 and 1919 respectively.

Sector mass spectrometers known as calutrons were used for separating the isotopes of uranium developed by Ernest O. Lawrence during the Manhattan Project. Calutron mass spectrometers were used for uranium enrichment at the Oak Ridge, Tennessee Y-12 plant established during World War II.

In 1989, half of the Nobel Prize in Physics was awarded to Hans Dehmelt and Wolfgang Paul for the development of the ion trap technique in the 1950s and 1960s.

In 2002, the Nobel Prize in Chemistry was awarded to John Bennett Fenn for the development of electrospray ionization (ESI) and Koichi Tanaka for the development of soft laser desorption (SLD) and their application to the ionization of biological macromolecules, especially proteins.

Parts of a Mass Spectrometer

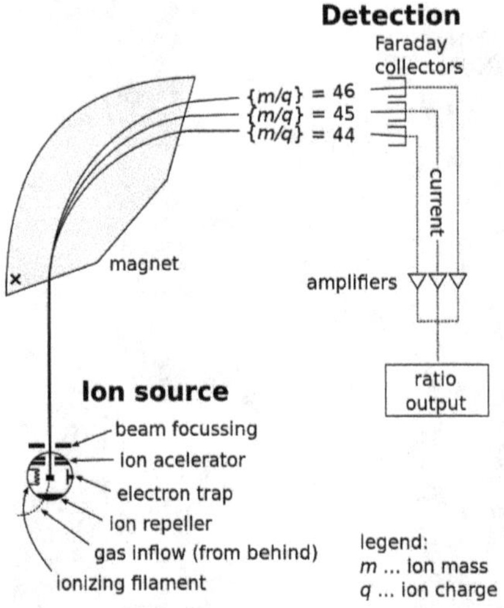

Schematics of a simple mass spectrometer with sector type mass analyzer. This one is for the measurement of carbon dioxide isotope ratios (IRMS) as in the carbon-13 urea breath test

A mass spectrometer consists of three components: an ion source, a mass analyzer, and a detector. The *ionizer* converts a portion of the sample into ions. There is a wide variety of ionization techniques, depending on the phase (solid, liquid, gas) of the sample and the efficiency of various ionization mechanisms for the unknown species. An extraction system removes ions from the sample, which are then targeted through the mass analyzer and onto the *detector*. The differences in masses of the fragments allows the mass analyzer to sort the ions by their mass-to-charge ratio. The detector measures the value of an indicator quantity and thus provides data for calculating the abundances of each ion present. Some detectors also give spatial information, e.g., a multichannel plate.

Theoretical Example

The following example describes the operation of a spectrometer mass analyzer, which is of the sector type. (Other analyzer types are treated below.) Consider a sample of sodium chloride (table salt). In the ion source, the sample is vaporized (turned into gas) and ionized (transformed into electrically charged particles) into sodium (Na^+) and chloride (Cl^-) ions. Sodium atoms and ions are monoisotopic, with a mass of about 23 u. Chloride atoms and ions come in two isotopes with masses of approximately 35 u (at a natural abundance of about 75 percent) and approximately 37 u (at a natural abundance of about 25 percent). The analyzer part of the spectrometer contains electric and magnetic fields, which exert forces on ions traveling through these fields. The speed of a charged particle may be increased or decreased while passing through the electric field, and its direction may be altered by the magnetic field. The magnitude of the deflection of the moving ion's trajectory depends on its mass-to-charge ratio. Lighter ions get deflected by the magnetic force more than heavier ions (based on Newton's second law of motion, $F = ma$). The streams of sorted ions pass from the analyzer to the detector, which records the relative abundance of each ion type. This information is used to determine the chemical element composition of the original sample (i.e. that both sodium and chlorine are present in the sample) and the isotopic composition of its constituents (the ratio of ^{35}Cl to ^{37}Cl).

Creating Ions

Surface ionization source at the Argonne National Laboratory linear accelerator

The ion source is the part of the mass spectrometer that ionizes the material under analysis (the analyte). The ions are then transported by magnetic or electric fields to the mass analyzer.

Techniques for ionization have been key to determining what types of samples can be analyzed by mass spectrometry. Electron ionization and chemical ionization are used for gases and vapors. In chemical ionization sources, the analyte is ionized by chemical ion-molecule reactions during collisions in the source. Two techniques often used with liquid and solid biological samples include electrospray ionization (invented by John Fenn) and matrix-assisted laser desorption/ionization (MALDI, initially developed as a similar technique "Soft Laser Desorption (SLD)" by K. Tanaka for which a Nobel Prize was awarded and as MALDI by M. Karas and F. Hillenkamp).

Hard Ionization and Soft Ionization

Quadrupole mass spectrometer and electrospray ion source used for Fenn's early work

In mass spectrometry, ionization refers to the production of gas phase ions suitable for resolution in the mass analyser or mass filter. Ionization occurs in the ion source. There are several ion sources available; each has advantages and disadvantages for particular applications. For example, electron ionization (EI) gives a high degree of fragmentation, yielding highly detailed mass spectra which when skilfully analysed can provide important information for structural elucidation/characterisation and facilitate identification of unknown compounds by comparison to mass spectral libraries obtained under identical operating conditions. However, EI is not suitable for coupling to HPLC, i.e. LC-MS, since at atmospheric pressure, the filaments used to generate electrons burn out rapidly. Thus EI is coupled predominantly with GC, i.e. GC-MS, where the entire system is under high vacuum.

Hard ionization techniques are processes which impart high quantities of residual energy in the subject molecule invoking large degrees of fragmentation (i.e. the systematic rupturing of bonds acts to remove the excess energy, restoring stability to the resulting ion). Resultant ions tend to have m/z lower than the molecular mass (other than in the case of proton transfer and not including isotope peaks). The most common example of hard ionization is electron ionization (EI).

Soft ionization refers to the processes which impart little residual energy onto the subject molecule and as such result in little fragmentation. Examples include fast atom bombardment (FAB), chemical ionization (CI), atmospheric-pressure chemical ionization (APCI), electrospray ionization (ESI), and matrix-assisted laser desorption/ionization (MALDI)

Inductively Coupled Plasma

Inductively coupled plasma (ICP) sources are used primarily for cation analysis of a wide array of sample types. In this source, a plasma that is electrically neutral overall, but that has had a substantial fraction of its atoms ionized by high temperature, is used to atomize introduced sample molecules and to further strip the outer electrons from those atoms. The plasma is usually generated from argon gas, since the first ionization energy of argon atoms is higher than the first of any other elements except He, O, F and Ne, but lower than the second ionization energy of all except the most electropositive metals. The heating is achieved by a radio-frequency current passed through a coil surrounding the plasma.

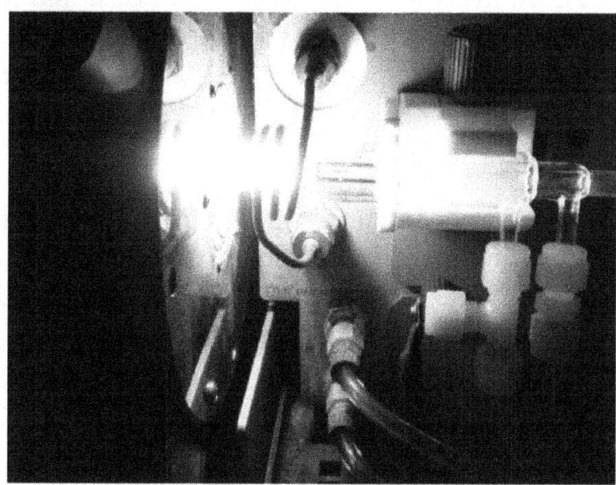

Inductively coupled plasma ion source

Other Ionization Techniques

Others include photoionization, glow discharge, field desorption (FD), fast atom bombardment (FAB), thermospray, desorption/ionization on silicon (DIOS), Direct Analysis in Real Time (DART), atmospheric pressure chemical ionization (APCI), secondary ion mass spectrometry (SIMS), spark ionization and thermal ionization (TIMS).

Mass Selection

Mass analyzers separate the ions according to their mass-to-charge ratio. The following two laws govern the dynamics of charged particles in electric and magnetic fields in vacuum:

$\mathbf{F} = Q(\mathbf{E} + \mathbf{v} \times \mathbf{B})$ (Lorentz force law);

$\mathbf{F} = m\mathbf{a}$ (Newton's second law of motion in non-relativistic case, i.e. valid only at ion velocity much lower than the speed of light).

Here F is the force applied to the ion, *m* is the mass of the ion, a is the acceleration, *Q* is the ion charge, E is the electric field, and v × B is the vector cross product of the ion velocity and the magnetic field

Equating the above expressions for the force applied to the ion yields:

$$(m / Q)\mathbf{a} = \mathbf{E} + \mathbf{v} \times \mathbf{B}.$$

This differential equation is the classic equation of motion for charged particles. Together with the particle's initial conditions, it completely determines the particle's motion in space and time in terms of m/Q. Thus mass spectrometers could be thought of as "mass-to-charge spectrometers". When presenting data, it is common to use the (officially) dimensionless m/z, where z is the number of elementary charges (e) on the ion (z=Q/e). This quantity, although it is informally called the mass-to-charge ratio, more accurately speaking represents the ratio of the mass number and the charge number, z.

There are many types of mass analyzers, using either static or dynamic fields, and magnetic or electric fields, but all operate according to the above differential equation. Each analyzer type has its strengths and weaknesses. Many mass spectrometers use two or more mass analyzers for tandem mass spectrometry (MS/MS). In addition to the more common mass analyzers listed below, there are others designed for special situations.

There are several important analyser characteristics. The mass resolving power is the measure of the ability to distinguish two peaks of slightly different m/z. The mass accuracy is the ratio of the m/z measurement error to the true m/z. Mass accuracy is usually measured in ppm or milli mass units. The mass range is the range of m/z amenable to analysis by a given analyzer. The linear dynamic range is the range over which ion signal is linear with analyte concentration. Speed refers to the time frame of the experiment and ultimately is used to determine the number of spectra per unit time that can be generated.

Sector Instruments

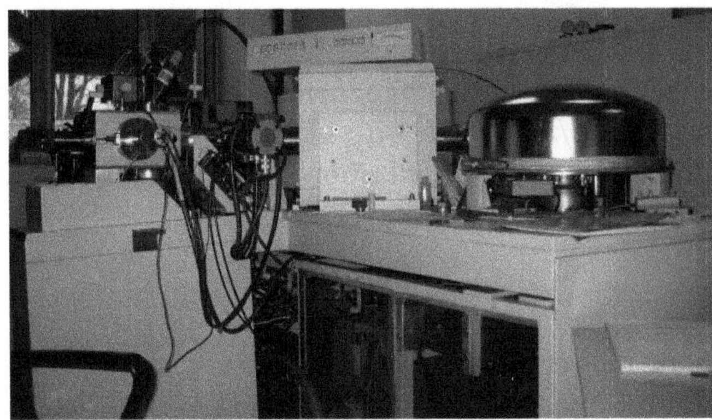

ThermoQuest AvantGarde sector mass spectrometer

A sector field mass analyzer uses an electric and/or magnetic field to affect the path and/or velocity of the charged particles in some way. As shown above, sector instruments bend the trajectories of the ions as they pass through the mass analyzer, according to their mass-to-charge ratios, deflecting the more charged and faster-moving, lighter ions more. The analyzer can be used to select a narrow range of m/z or to scan through a range of m/z to catalog the ions present.

Time-of-flight

The time-of-flight (TOF) analyzer uses an electric field to accelerate the ions through the same potential, and then measures the time they take to reach the detector. If the particles all have the same charge, the kinetic energies will be identical, and their velocities will depend only on their masses. Ions with a lower mass will reach the detector first.

Quadrupole Mass Filter

Quadrupole mass analyzers use oscillating electrical fields to selectively stabilize or destabilize the paths of ions passing through a radio frequency (RF) quadrupole field created between 4 parallel rods. Only the ions in a certain range of mass/charge ratio are passed through the system at any time, but changes to the potentials on the rods allow a wide range of m/z values to be swept rapidly, either continuously or in a succession of discrete hops. A quadrupole mass analyzer acts as a mass-selective filter and is closely related to the quadrupole ion trap, particularly the linear quadrupole ion trap except that it is designed to pass the untrapped ions rather than collect the trapped ones, and is for that reason referred to as a transmission quadrupole. A magnetically enhanced quadrupole mass analyzer includes the addition of a magnetic field, either applied axially or transversely. This novel type of instrument leads to an additional performance enhancement in terms of resolution and/or sensitivity depending upon the magnitude and orientation of the applied magnetic field. A common variation of the transmission quadrupole is the triple quadrupole mass spectrometer. The "triple quad" has three consecutive quadrupole stages, the first acting as a mass filter to transmit a particular incoming ion to the second quadrupole, a collision chamber, wherein that ion can be broken into fragments. The third quadrupole also acts as a mass filter, to transmit a particular fragment ion to the detector. If a quadrupole is made to rapidly and repetitively cycle through a range of mass filter settings, full spectra can be reported. Likewise, a triple quad can be made to perform various scan types characteristic of tandem mass spectrometry.

Ion traps

Three-dimensional Quadrupole Ion Trap

The quadrupole ion trap works on the same physical principles as the quadrupole mass analyzer, but the ions are trapped and sequentially ejected. Ions are trapped in a mainly quadrupole RF field, in a space defined by a ring electrode (usually connected to the main RF potential) between two endcap electrodes (typically connected to DC or auxiliary AC potentials). The sample is ionized either internally (e.g. with an electron or laser beam), or externally, in which case the ions are often introduced through an aperture in an endcap electrode.

There are many mass/charge separation and isolation methods but the most commonly used is the mass instability mode in which the RF potential is ramped so that the orbit of ions with a mass $a > b$ are stable while ions with mass b become unstable and are ejected on the z-axis onto a detector. There are also non-destructive analysis methods.

Ions may also be ejected by the resonance excitation method, whereby a supplemental oscillatory excitation voltage is applied to the endcap electrodes, and the trapping voltage amplitude and/or excitation voltage frequency is varied to bring ions into a resonance condition in order of their mass/charge ratio.

Cylindrical Ion Trap

The cylindrical ion trap mass spectrometer (CIT) is a derivative of the quadrupole ion trap where the electrodes are formed from flat rings rather than hyperbolic shaped electrodes. The architecture lends itself well to miniaturization because as the size of a trap is reduced, the shape of the electric field near the center of the trap, the region where the ions are trapped, forms a shape similar to that of a hyperbolic trap.

Linear Quadrupole Ion Trap

A linear quadrupole ion trap is similar to a quadrupole ion trap, but it traps ions in a two dimensional quadrupole field, instead of a three-dimensional quadrupole field as in a 3D quadrupole ion trap. Thermo Fisher's LTQ ("linear trap quadrupole") is an example of the linear ion trap.

A toroidal ion trap can be visualized as a linear quadrupole curved around and connected at the ends or as a cross section of a 3D ion trap rotated on edge to form the toroid, donut shaped trap. The trap can store large volumes of ions by distributing them throughout the ring-like trap structure. This toroidal shaped trap is a configuration that allows the increased miniaturization of an ion trap mass analyzer. Additionally all ions are stored in the same trapping field and ejected together simplifying detection that can be complicated with array configurations due to variations in detector alignment and machining of the arrays.

As with the toroidal trap, linear traps and 3D quadrupole ion traps are the most commonly miniaturized mass analyzers due to their high sensitivity, tolerance for mTorr pressure, and capabilities for single analyzer tandem mass spectrometry (e.g. product ion scans).

Orbitrap

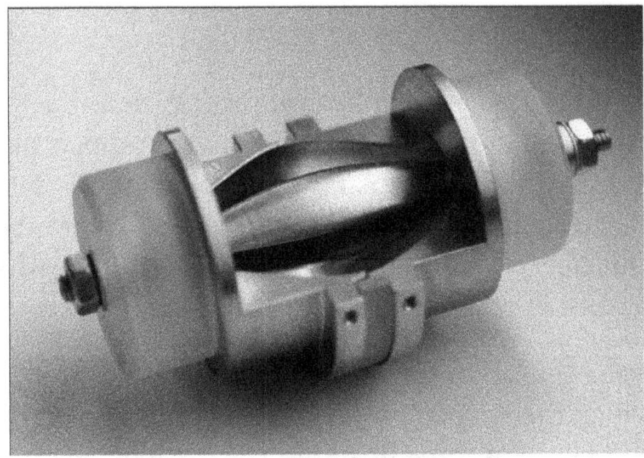

Orbitrap mass analyzer

Orbitrap instruments are similar to Fourier transform ion cyclotron resonance mass spectrometers. Ions are electrostatically trapped in an orbit around a central, spindle shaped electrode. The electrode confines the ions so that they both orbit around the central electrode and oscillate back and forth along the central electrode's long axis. This oscillation generates an image current in the detector plates which is recorded by the instrument. The frequencies of these image currents depend on the mass to charge ratios of the ions. Mass spectra are obtained by Fourier transformation of the recorded image currents.

Orbitraps have a high mass accuracy, high sensitivity and a good dynamic range.

Fourier Transform Ion Cyclotron Resonance

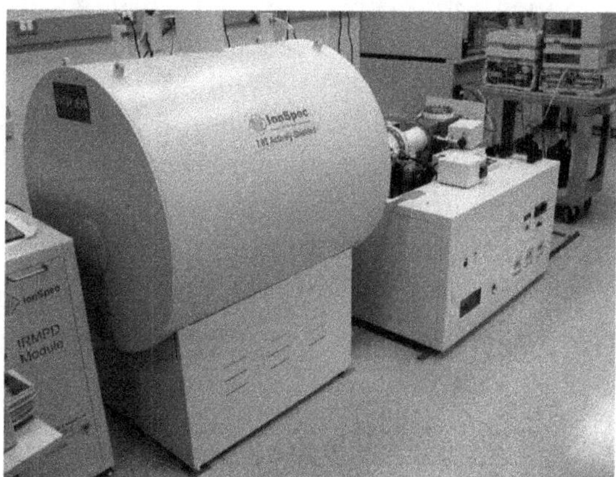

Fourier transform ion cyclotron resonance mass spectrometer

Fourier transform mass spectrometry (FTMS), or more precisely Fourier transform ion cyclotron resonance MS, measures mass by detecting the image current produced by ions cyclotroning in the presence of a magnetic field. Instead of measuring the deflection of ions with a detector such as an electron multiplier, the ions are injected into a Penning trap (a static electric/magnetic ion trap) where they effectively form part of a circuit. Detectors at fixed positions in space measure the electrical signal of ions which pass near them over time, producing a periodic signal. Since the frequency of an ion's cycling is determined by its mass to charge ratio, this can be deconvoluted by performing a Fourier transform on the signal. FTMS has the advantage of high sensitivity (since each ion is "counted" more than once) and much higher resolution and thus precision.

Ion cyclotron resonance (ICR) is an older mass analysis technique similar to FTMS except that ions are detected with a traditional detector. Ions trapped in a Penning trap are excited by an RF electric field until they impact the wall of the trap, where the detector is located. Ions of different mass are resolved according to impact time.

Detectors

A continuous dynode particle multiplier detector.

The final element of the mass spectrometer is the detector. The detector records either the charge induced or the current produced when an ion passes by or hits a surface. In a scanning instrument, the signal produced in the detector during the course of the scan versus where the instrument is in the scan (at what m/Q) will produce a mass spectrum, a record of ions as a function of m/Q.

Typically, some type of electron multiplier is used, though other detectors including Faraday cups and ion-to-photon detectors are also used. Because the number of ions leaving the mass analyzer at a particular instant is typically quite small, considerable amplification is often necessary to get a signal. Microchannel plate detectors are commonly used in modern commercial instruments. In FTMS and Orbitraps, the detector consists of a pair of metal surfaces within the mass analyzer/ion trap region which the ions only pass near as they oscillate. No direct current is produced, only a weak AC image current is produced in a circuit between the electrodes. Other inductive detectors have also been used.

Tandem Mass Spectrometry

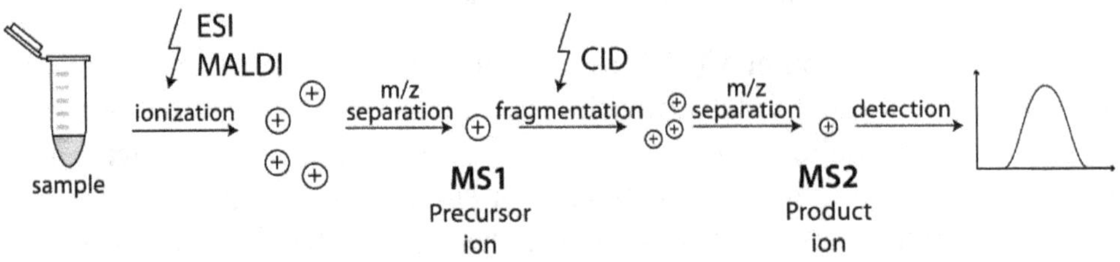

Tandem mass spectrometry for biological molecules using ESI or MALDI

A tandem mass spectrometer is one capable of multiple rounds of mass spectrometry, usually separated by some form of molecule fragmentation. For example, one mass analyzer can isolate one peptide from many entering a mass spectrometer. A second mass analyzer then stabilizes the peptide ions while they collide with a gas, causing them to fragment by collision-induced dissociation (CID). A third mass analyzer then sorts the fragments produced from the peptides. Tandem MS can also be done in a single mass analyzer over time, as in a quadrupole ion trap. There are various methods for fragmenting molecules for tandem MS, including collision-induced dissociation (CID), electron capture dissociation (ECD), electron transfer dissociation (ETD), infrared multiphoton dissociation (IRMPD), blackbody infrared radiative dissociation (BIRD), electron-detachment dissociation (EDD) and surface-induced dissociation (SID). An important application using tandem mass spectrometry is in protein identification.

Tandem mass spectrometry enables a variety of experimental sequences. Many commercial mass spectrometers are designed to expedite the execution of such routine sequences as selected reaction monitoring (SRM) and precursor ion scanning. In SRM, the first analyzer allows only a single mass through and the second analyzer monitors for multiple user-defined fragment ions. SRM is most often used with scanning instruments where the second mass analysis event is duty cycle limited. These experiments are used to increase specificity of detection of known molecules, notably in pharmacokinetic studies. Precursor ion scanning refers to monitoring for a specific loss from the precursor ion. The first and second mass analyzers scan across the spectrum as partitioned by a user-defined m/z value. This experiment is used to detect specific motifs within unknown molecules.

Another type of tandem mass spectrometry used for radiocarbon dating is accelerator mass spectrometry (AMS), which uses very high voltages, usually in the mega-volt range, to accelerate negative ions into a type of tandem mass spectrometer.

Common Mass Spectrometer Configurations and Techniques

When a specific configuration of source, analyzer, and detector becomes conventional in practice, often a compound acronym arises to designate it, and the compound acronym may be better known among nonspectrometrists than the component acronyms. The epitome of this is MALDI-TOF, which simply refers to combining a matrix-assisted laser desorption/ionization source with a time-of-flight mass analyzer. The MALDI-TOF moniker is more widely recognized by the non-mass spectrometrists than MALDI or TOF individually. Other examples include inductively coupled plasma-mass spectrometry (ICP-MS), accelerator mass spectrometry (AMS), thermal ionization-mass spectrometry (TIMS) and spark source mass spectrometry (SSMS). Sometimes the use of the generic "MS" actually connotes a very specific mass analyzer and detection system, as is the case with AMS, which is always sector based.

Certain applications of mass spectrometry have developed monikers that although strictly speaking would seem to refer to a broad application, in practice have come instead to connote a specific or a limited number of instrument configurations. An example of this is isotope ratio mass spectrometry (IRMS), which refers in practice to the use of a limited number of sector based mass analyzers; this name is used to refer to both the application and the instrument used for the application.

Separation Techniques Combined With Mass Spectrometry

An important enhancement to the mass resolving and mass determining capabilities of mass spectrometry is using it in tandem with chromatographic and other separation techniques.

Gas Chromatography

A gas chromatograph (right) directly coupled to a mass spectrometer (left)

A common combination is gas chromatography-mass spectrometry (GC/MS or GC-MS). In this technique, a gas chromatograph is used to separate different compounds. This stream of separated compounds is fed online into the ion source, a metallic filament to which voltage is applied. This filament emits electrons which ionize the compounds. The ions can then further fragment, yielding predictable patterns. Intact ions and fragments pass into the mass spectrometer's analyzer and are eventually detected.

Liquid Chromatography

Similar to gas chromatography MS (GC/MS), liquid chromatography-mass spectrometry (LC/MS or LC-MS) separates compounds chromatographically before they are introduced to the ion source and mass spectrometer. It differs from GC/MS in that the mobile phase is liquid, usually a mixture of water and organic solvents, instead of gas. Most commonly, an electrospray ionization source is used in LC/MS. Other popular and commercially available LC/MS ion sources are atmospheric pressure chemical ionization and atmospheric pressure photoionization. There are also some newly developed ionization techniques like laser spray.

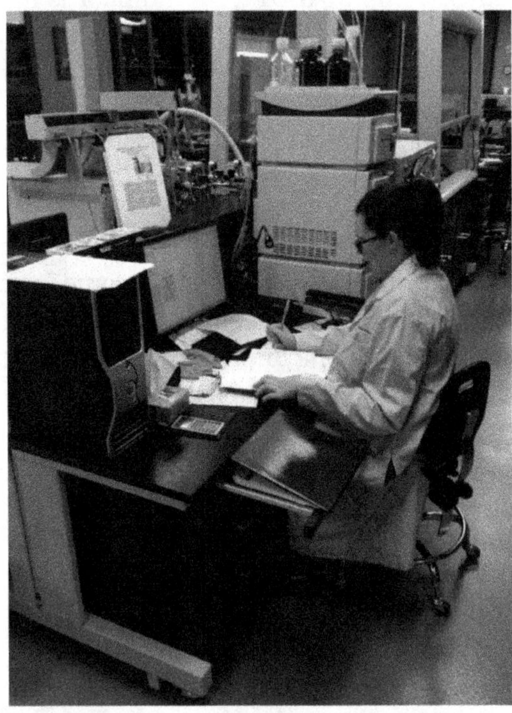

Indianapolis Museum of Art conservation scientist performing liquid chromatography–mass spectrometry.

Capillary Electrophoresis–mass Spectrometry

Capillary electrophoresis–mass spectrometry (CE-MS) is a technique that combines the liquid separation process of capillary electrophoresis with mass spectrometry. CE-MS is typically coupled to electrospray ionization.

Ion Mobility

Ion mobility spectrometry-mass spectrometry (IMS/MS or IMMS) is a technique where ions are first separated by drift time through some neutral gas under an applied electrical potential gradient before being introduced into a mass spectrometer. Drift time is a measure of the radius relative to the charge of the ion. The duty cycle of IMS (the time over which the experiment takes place) is longer than most mass spectrometric techniques, such that the mass spectrometer can sample along the course of the IMS separation. This produces data about the IMS separation and the mass-to-charge ratio of the ions in a manner similar to LC/MS.

The duty cycle of IMS is short relative to liquid chromatography or gas chromatography separations and can thus be coupled to such techniques, producing triple modalities such as LC/IMS/MS.

Data and Analysis

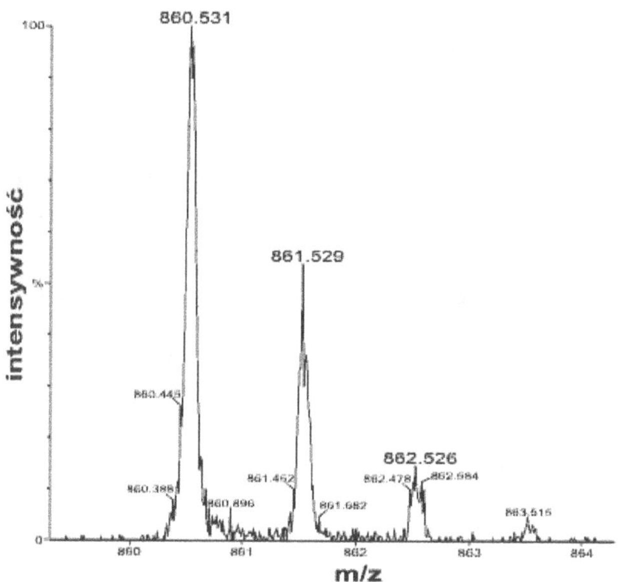

Mass spectrum of a peptide showing the isotopic distribution

Data Representations

Mass spectrometry produces various types of data. The most common data representation is the mass spectrum.

Certain types of mass spectrometry data are best represented as a mass chromatogram. Types of chromatograms include selected ion monitoring (SIM), total ion current (TIC), and selected reaction monitoring (SRM), among many others.

Other types of mass spectrometry data are well represented as a three-dimensional contour map. In this form, the mass-to-charge, m/z is on the x-axis, intensity the y-axis, and an additional experimental parameter, such as time, is recorded on the z-axis.

Data Analysis

Mass spectrometry data analysis is specific to the type of experiment producing the data. General subdivisions of data are fundamental to understanding any data.

Many mass spectrometers work in either *negative ion mode* or *positive ion mode*. It is very important to know whether the observed ions are negatively or positively charged. This is often important in determining the neutral mass but it also indicates something about the nature of the molecules.

Different types of ion source result in different arrays of fragments produced from the original molecules. An electron ionization source produces many fragments and mostly single-charged (1-) radicals (odd number of electrons), whereas an electrospray source usually produces non-radical quasimolecular ions that are frequently multiply charged. Tandem mass spectrometry purposely produces fragment ions post-source and can drastically change the sort of data achieved by an experiment.

Knowledge of the origin of a sample can provide insight into the component molecules of the sample and their fragmentations. A sample from a synthesis/manufacturing process will probably contain impurities chemically related to the target component. A crudely prepared biological sample will probably contain a certain amount of salt, which may form adducts with the analyte molecules in certain analyses.

Results can also depend heavily on sample preparation and how it was run/introduced. An important example is the issue of which matrix is used for MALDI spotting, since much of the energetics of the desorption/ionization event is controlled by the matrix rather than the laser power. Sometimes samples are spiked with sodium or another ion-carrying species to produce adducts rather than a protonated species.

Mass spectrometry can measure molar mass, molecular structure, and sample purity. Each of these questions requires a different experimental procedure; therefore, adequate definition of the experimental goal is a prerequisite for collecting the proper data and successfully interpreting it.

Interpretation of Mass Spectra

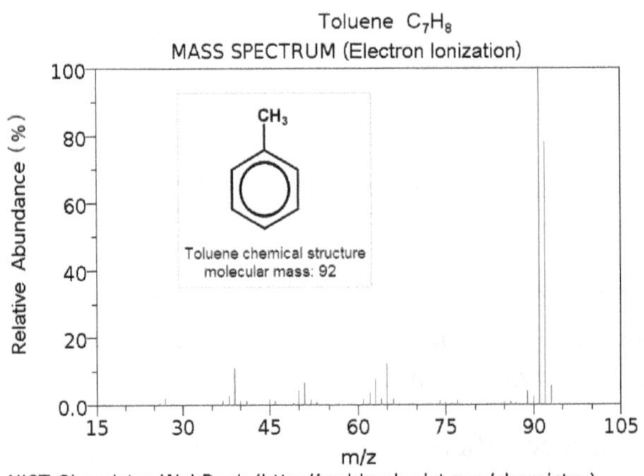

Toluene electron ionization mass spectrum

Since the precise structure or peptide sequence of a molecule is deciphered through the set of fragment masses, the interpretation of mass spectra requires combined use of various techniques. Usually the first strategy for identifying an unknown compound is to compare its experimental mass spectrum against a library of mass spectra. If no matches result from the search, then manual interpretation or software assisted interpretation of mass spectra must be performed. Computer simulation of ionization and fragmentation processes occurring in mass spectrometer is the primary tool for assigning structure or peptide sequence to a molecule. An *a priori* structural information is fragmented *in silico* and the resulting pattern is compared with observed spectrum. Such simulation is often supported by a fragmentation library that contains published patterns of known decomposition reactions. Software taking advantage of this idea has been developed for both small molecules and proteins.

Analysis of mass spectra can also be spectra with accurate mass. A mass-to-charge ratio value (m/z) with only integer precision can represent an immense number of theoretically possible ion structures; however, more precise mass figures significantly reduce the number of candidate molecular formulas. A computer algorithm called formula generator calculates all molecular formulas that theoretically fit a given mass with specified tolerance.

A recent technique for structure elucidation in mass spectrometry, called precursor ion fingerprinting, identifies individual pieces of structural information by conducting a search of the tandem spectra of the molecule under investigation against a library of the product-ion spectra of structurally characterized precursor ions.

Applications

NOAA Particle Analysis by Laser Mass Spectrometry aerosol mass spectrometer aboard a NASA WB-57 high-altitude research aircraft

Mass spectrometry has both qualitative and quantitative uses. These include identifying unknown compounds, determining the isotopic composition of elements in a molecule, and determining the structure of a compound by observing its fragmentation. Other uses include quantifying the amount of a compound in a sample or studying the fundamentals of gas phase ion chemistry (the chemistry of ions and neutrals in a vacuum). MS is now in very common use in analytical laboratories that study physical, chemical, or biological properties of a great variety of compounds.

As an analytical technique it possesses distinct advantages such as: Increased sensitivity over most other analytical techniques because the analyzer, as a mass-charge filter, reduces background interference, Excellent specificity from characteristic fragmentation patterns to identify unknowns or confirm the presence of suspected compounds, Information about molecular weight, Information about the isotopic abundance of elements, Temporally resolved chemical data.

A few of the disadvantages of the method is that often fails to distinguish between optical and geometrical isomers and the positions of substituent in o-, m- and p- positions in an aromatic ring. Also, its scope is limited in identifying hydrocarbons that produce similar fragmented ions.

Isotope Ratio MS: Isotope Dating and Tracing

Mass spectrometry is also used to determine the isotopic composition of elements within a sample. Differences in mass among isotopes of an element are very small, and the less abundant isotopes of an element are typically very rare, so a very sensitive instrument is required. These instruments, sometimes referred to as isotope ratio mass spectrometers (IR-MS), usually use a single magnet to bend a beam of ionized particles towards a series of Faraday cups which convert particle impacts to electric current. A fast on-line analysis of deuterium content of water can be done using flowing afterglow mass spectrometry, FA-MS. Probably the most sensitive and accurate mass spectrometer for this purpose is the accelerator mass spectrometer (AMS). This is because it provides ultimate sensitivity, capable of measuring individual atoms and measuring nuclides with a dynamic range of ~10^{15} relative to the major stable isotope. Isotope ratios are important markers of a variety of processes. Some isotope ratios are used to determine the age of materials for example as in carbon dating. Labeling with stable isotopes is also used for protein quantification.

Mass spectrometer to determine the $^{16}O/^{18}O$ and $^{12}C/^{13}C$ isotope ratio on biogenous carbonate

Trace Gas Analysis

Several techniques use ions created in a dedicated ion source injected into a flow tube or a drift tube: selected ion flow tube (SIFT-MS), and proton transfer reaction (PTR-MS), are variants of chemical ionization dedicated for trace gas analysis of air, breath or liquid headspace using well defined reaction time allowing calculations of analyte concentrations from the known reaction kinetics without the need for internal standard or calibration.

Atom Probe

An atom probe is an instrument that combines time-of-flight mass spectrometry and field-evaporation microscopy to map the location of individual atoms.

Pharmacokinetics

Pharmacokinetics is often studied using mass spectrometry because of the complex nature of the matrix (often blood or urine) and the need for high sensitivity to observe low dose and long time point data. The most common instrumentation used in this application is LC-MS with a triple quadrupole mass spectrometer. Tandem mass spectrometry is usually employed for added specificity. Standard curves and internal standards are used for quantitation of usually a single pharmaceutical in the samples. The samples represent different time points as a pharmaceutical is administered and then metabolized or cleared from the body. Blank or t=0 samples taken before administration are important in determining background and ensuring data integrity with such complex sample matrices. Much attention is paid to the linearity of the standard curve; however it is not uncommon to use curve fitting with more complex functions such as quadratics since the response of most mass spectrometers is less than linear across large concentration ranges.

There is currently considerable interest in the use of very high sensitivity mass spectrometry for microdosing studies, which are seen as a promising alternative to animal experimentation.

Protein Characterization

Mass spectrometry is an important method for the characterization and sequencing of proteins. The two primary methods for ionization of whole proteins are electrospray ionization (ESI) and matrix-assisted laser desorption/ionization (MALDI). In keeping with the performance and mass range of available mass spectrometers, two approaches are used for characterizing proteins. In the first, intact proteins are ionized by either of the two techniques described above, and then introduced to a mass analyzer. This approach is referred to as "top-down" strategy of protein analysis. In the second, proteins are enzymatically digested into smaller peptides using proteases such as trypsin or pepsin, either in solution or in gel after electrophoretic separation. Other proteolytic agents are also used. The collection of peptide products are then introduced to the mass analyzer. When the characteristic pattern of peptides is used for the identification of the protein the method is called peptide mass fingerprinting (PMF), if the identification is performed using the sequence data determined in tandem MS analysis it is called de novo peptide sequencing. These procedures of protein analysis are also referred to as the "bottom-up" approach.

Glycan Analysis

Mass spectrometry (MS), with its low sample requirement and high sensitivity, has been predominantly used in glycobiology for characterization and elucidation of glycan structures. Mass spectrometry provides a complementary method to HPLC for the analysis of glycans. Intact glycans may be detected directly as singly charged ions by matrix-assisted laser desorption/ionization mass spectrometry (MALDI-MS) or, following permethylation or peracetylation, by fast atom bombardment mass spectrometry (FAB-MS). Electrospray ionization mass spectrometry (ESI-MS) also gives good signals for the smaller glycans. Various free and commercial software are now available which interpret MS data and aid in Glycan structure characterization.

Space Exploration

As a standard method for analysis, mass spectrometers have reached other planets and moons. Two were taken to Mars by the Viking program. In early 2005 the Cassini–Huygens mission delivered a specialized GC-MS instrument aboard the Huygens probe through the atmosphere of Titan, the largest moon of the planet Saturn. This instrument analyzed atmospheric samples along its descent trajectory and was able to vaporize and analyze samples of Titan's frozen, hydrocarbon covered surface once the probe had landed. These measurements compare the abundance of isotope(s) of each particle comparatively to earth's natural abundance. Also on board the Cassini–Huygens spacecraft is an ion and neutral mass spectrometer which has been taking measurements of Titan's atmospheric composition as well as the composition of Enceladus' plumes. A Thermal and Evolved Gas Analyzer mass spectrometer was carried by the Mars Phoenix Lander launched in 2007.

NASA's Phoenix Mars Lander analyzing a soil sample from the "Rosy Red" trench with the TEGA mass spectrometer

Mass spectrometers are also widely used in space missions to measure the composition of plasmas. For example, the Cassini spacecraft carries the Cassini Plasma Spectrometer (CAPS), which measures the mass of ions in Saturn's magnetosphere.

Respired Gas Monitor

Mass spectrometers were used in hospitals for respiratory gas analysis beginning around 1975 through the end of the century. Some are probably still in use but none are currently being manufactured.

Found mostly in the operating room, they were a part of a complex system, in which respired gas samples from patients undergoing anesthesia were drawn into the instrument through a valve mechanism designed to sequentially connect up to 32 rooms to the mass spectrometer. A computer directed all operations of the system. The data collected from the mass spectrometer was delivered to the individual rooms for the anesthesiologist to use.

The uniqueness of this magnetic sector mass spectrometer may have been the fact that a plane of detectors, each purposely positioned to collect all of the ion species expected to be in the samples, allowed the instrument to simultaneously report all of the gases respired by the patient. Although the mass range was limited to slightly over 120 u, fragmentation of some of the heavier molecules negated the need for a higher detection limit.

Preparative Mass Spectrometry

The primary function of mass spectrometry is as a tool for chemical analyses based on detection and quantification of ions according to their mass-to-charge ratio. However, mass spectrometry also shows promise for material synthesis. Ion soft landing is characterized by deposition of intact species on surfaces at low kinetic energies which precludes the fragmentation of the incident species. The soft landing technique was first reported in 1977 for the reaction of low energy sulfur containing ions on a lead surface.

Peptide Mass Fingerprinting

Peptide mass fingerprinting (PMF) (also known as protein fingerprinting) is an analytical technique for protein identification in which the unknown protein of interest is first cleaved into smaller peptides, whose absolute masses can be accurately measured with a mass spectrometer such as MALDI-TOF or ESI-TOF. The method was developed in 1993 by several groups independently. The peptide masses are compared to either a database containing known protein sequences or even the genome. This is achieved by using computer programs that translate the known genome of the organism into proteins, then theoretically cut the proteins into peptides, and calculate the absolute masses of the peptides from each protein. They then compare the masses of the peptides of the unknown protein to the theoretical peptide masses of each protein encoded in the genome. The results are statistically analyzed to find the best match. The advantage of this method is that only the masses of the peptides have to be known. Time-consuming de novo peptide sequencing is then unnecessary. A disadvantage is that the protein sequence has to be present in the database of interest. Additionally most PMF algorithms assume that the peptides come from a single protein. The presence of a mixture can significantly complicate the analysis and potentially compromise the results. Typical for the PMF based protein identification is the requirement for an isolated protein. Mixtures exceeding a number of 2-3 proteins typically require the additional use of MS/MS based protein identification to achieve sufficient specificity of identification (6). Therefore, the typical PMF samples are isolated proteins from two-dimensional gel electrophoresis (2D gels) or isolated SDS-PAGE bands. Additional analyses by MS/MS can either be direct, e.g., MALDI-TOF/TOF analysis or downstream nanoLC-ESI-MS/MS analysis of gel spot eluates.

Sample Preparation

Protein samples can be derived from SDS-PAGE and are then subject to some chemical modifications. Disulfide bridges in proteins are reduced and cysteine amino acids are carbamidomethylated chemically or acrylamidated during the gel electrophoresis.

Then the proteins are cut into several fragments using proteolytic enzymes such as trypsin, chymotrypsin or Glu-C. A typical sample:protease ratio is 50:1. The proteolysis is typically carried out overnight and the resulting peptides are extracted with acetonitrile and dried under vacuum. The peptides are then dissolved in a small amount of distilled water or further concentrated and purified using ZipTip Pipette tips and are ready for mass spectrometric analysis.

Mass Spectrometric Analysis

The digested protein can be analyzed with different types of mass spectrometers such as ESI-TOF or MALDI-TOF. MALDI-TOF is often the preferred instrument because it allows a high sample throughput and several proteins can be analyzed in a single experiment, if complemented by MS/MS analysis.

A small fraction of the peptide (usually 1 microliter or less) is pipetted onto a MALDI target and a chemical called a matrix is added to the peptide mix. The matrix molecules are required for the desorption of the peptide molecules. Matrix and peptide molecules co-crystallize on the MALDI target and are ready to be analyzed.

The target is inserted into the vacuum chamber of the mass spectrometer and the desorption and ionisation of the polypeptide fragments is initiated by a pulsed laser beam which transfers high amounts of energy into the matrix molecules. The energy transfer is sufficient to promote the ionisation and transition of matrix molecules and peptides from the solid phase into the gas phase. The ions are accelerated in the electric field of the mass spectrometer and fly towards an ion detector where their arrival is detected as an electric signal. Their mass-to-charge ratio is proportional to their time of flight (TOF) in the drift tube and can be calculated accordingly.

Computational Analysis

The mass spectrometric analysis produces a list of molecular weights of the fragments which is often called a peak list. The peptide masses are compared to protein databases such as Swissprot, which contain protein sequence information. Software performs *in silico* digests on proteins in the database with the same enzyme (e.g. trypsin) used in the chemical cleavage reaction. The mass of these peptide fragments is then calculated and compared to the peak list of measured peptide masses. The results are statistically analyzed and possible matches are returned in a results table.

Immunoprecipitation

Immunoprecipitation (IP) is the technique of precipitating a protein antigen out of solution using an antibody that specifically binds to that particular protein. This process can be used to isolate and concentrate a particular protein from a sample containing many thousands of different proteins. Immunoprecipitation requires that the antibody be coupled to a solid substrate at some point in the procedure.

Types of Immunoprecipitation

Individual Protein Immunoprecipitation (IP)

Involves using an antibody that is specific for a known protein to isolate that particular protein out of a solution containing many different proteins. These solutions will often be in the form of a crude lysate of a plant or animal tissue. Other sample types could be body fluids or other samples of biological origin.

Protein Complex Immunoprecipitation (Co-IP)

Immunoprecipitation of intact protein complexes (i.e. antigen along with any proteins or ligands that are bound to it) is known as co-immunoprecipitation (Co-IP). Co-IP works by selecting an antibody that targets a known protein that is believed to be a member of a larger complex of proteins. By targeting this *known* member with an antibody it may become possible to pull the entire protein complex out of solution and thereby identify *unknown* members of the complex.

This works when the proteins involved in the complex bind to each other tightly, making it possible to pull multiple members of the complex out of solution by latching onto one member with an antibody. This concept of pulling protein complexes out of solution is sometimes referred to as a "pull-down". Co-IP is a powerful technique that is used regularly by molecular biologists to analyze protein–protein interactions.

- A particular antibody often selects for a subpopulation of its target protein that has the epitope exposed, thus failing to identify any proteins in complexes that hide the epitope. This can be seen in that it is rarely possible to precipitate even half of a given protein from a sample with a single antibody, even when a large excess of antibody is used.
- The first round that were not identified in the previous experiment. As successive rounds of targeting and immunoprecipitations take place, the number of identified proteins may continue to grow. The identified proteins may not ever exist in a single complex at a given time, but may instead represent a network of proteins interacting with one another at different times for different purposes.
- Repeating the experiment by targeting different members of the protein complex allows the researcher to double-check the result. Each round of pull-downs should result in the recovery of both the original known protein as well as other previously identified members of the complex (and even new additional members). By repeating the immunoprecipitation in this way, the researcher verifies that each identified member of the protein complex was a valid identification. If a particular protein can only be recovered by targeting one of the known members but not by targeting other of the known members then that protein's status as a member of the complex may be subject to question.

Chromatin Immunoprecipitation (ChIP)

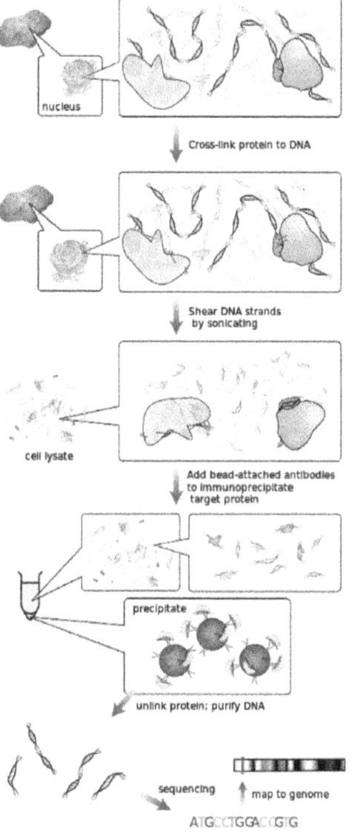

ChIP-sequencing workflow

Chromatin immunoprecipitation (ChIP) is a method used to determine the location of DNA binding sites on the genome for a particular protein of interest. This technique gives a picture of the protein–DNA interactions that occur inside the nucleus of living cells or tissues. The *in vivo* nature of this method is in contrast to other approaches traditionally employed to answer the same questions.

The principle underpinning this assay is that DNA-binding proteins (including transcription factors and histones) in living cells can be cross-linked to the DNA that they are binding. By using an antibody that is specific to a putative DNA binding protein, one can immunoprecipitate the protein–DNA complex out of cellular lysates. The crosslinking is often accomplished by applying formaldehyde to the cells (or tissue), although it is sometimes advantageous to use a more defined and consistent crosslinker such as DTBP. Following crosslinking, the cells are lysed and the DNA is broken into pieces 0.2–1.0 kb in length by sonication. At this point the immunoprecipitation is performed resulting in the purification of protein–DNA complexes. The purified protein–DNA complexes are then heated to reverse the formaldehyde cross-linking of the protein and DNA complexes, allowing the DNA to be separated from the proteins. The identity and quantity of the DNA fragments isolated can then be determined by PCR. The limitation of performing PCR on the isolated fragments is that one must have an idea which genomic region is being targeted in order to generate the correct PCR primers. This limitation is very easily circumvented simply by cloning the isolated genomic DNA into a plasmid vector and then using primers that are specific to the cloning region of that vector. Alternatively, when one wants to find where the protein binds on a genome-wide scale, a DNA microarray can be used (ChIP-on-chip or ChIP-chip) allowing for the characterization of the cistrome. As well, ChIP-Sequencing has recently emerged as a new technology that can localize protein binding sites in a high-throughput, cost-effective fashion.

RNA Immunoprecipitation (RIP)

Similar to chromatin immunoprecipitation (ChIP) outlined above, but rather than targeting DNA binding proteins as in ChIP, RNA immunoprecipitation targets RNA binding proteins. Live cells are first lysed and then the target protein and associated RNA are immunoprecipitated using an antibody targeting the protein of interest. The purified RNA-protein complexes can be separated by performing an RNA extraction and the identity of the RNA can be determined by cDNA sequencing or RT-PCR. Some variants of RIP, such as PAR-CLIP include cross-linking steps, which then require less careful lysis conditions.

Tagged Proteins

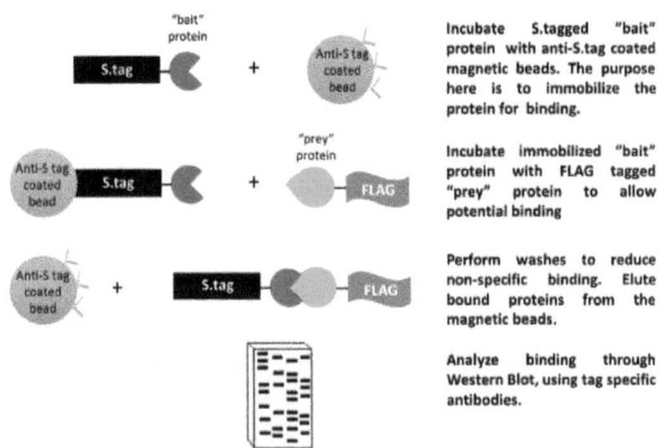

Pull down assay using tagged proteins

One of the major technical hurdles with immunoprecipitation is the great difficulty in generating an antibody that specifically targets a single known protein. To get around this obstacle, many groups will engineer tags onto either the C- or N- terminal end of the protein of interest. The advantage here is that the same tag can be used time and again on many different proteins and the researcher can use the same antibody each time. The advantages with using tagged proteins are so great that this technique has become commonplace for all types of immunoprecipitation including all of the types of IP detailed above. Examples of tags in use are the Green Fluorescent Protein (GFP) tag, Glutathione-S-transferase (GST) tag and the FLAG-tag tag. While the use of a tag to enable pull-downs is convenient, it raises some concerns regarding biological relevance because the tag itself may either obscure native interactions or introduce new and unnatural interactions.

Methods

The two general methods for immunoprecipitation are the direct capture method and the indirect capture method.

Direct

Antibodies that are specific for a particular protein (or group of proteins) are immobilized on a solid-phase substrate such as superparamagnetic microbeads or on microscopic agarose (non-magnetic) beads. The beads with bound antibodies are then added to the protein mixture, and the proteins that are targeted by the antibodies are captured onto the beads via the antibodies; in other words, they become immunoprecipitated.

Indirect

Antibodies that are specific for a particular protein, or a group of proteins, are added directly to the mixture of protein. The antibodies have not been attached to a solid-phase support yet. The antibodies are free to float around the protein mixture and bind their targets. As time passes, the beads coated in protein A/G are added to the mixture of antibody and protein. At this point, the antibodies, which are now bound to their targets, will stick to the beads.

From this point on, the direct and indirect protocols converge because the samples now have the same ingredients. Both methods gives the same end-result with the protein or protein complexes bound to the antibodies which themselves are immobilized onto the beads.

Selection

An indirect approach is sometimes preferred when the concentration of the protein target is low or when the specific affinity of the antibody for the protein is weak. The indirect method is also used when the binding kinetics of the antibody to the protein is slow for a variety of reasons. In most situations, the direct method is the default, and the preferred, choice.

Technological Advances

Agarose

Historically the solid-phase support for immunoprecipitation used by the majority of scientists has been highly-porous agarose beads (also known as agarose resins or slurries). The advantage of this technology is a very high potential binding capacity, as virtually the entire sponge-like structure of the agarose particle (50 to 150μm in size) is available for binding antibodies (which will in turn bind the target proteins) and the use of standard laboratory equipment for all aspects of the IP protocol without the need for any specialized equipment. The advantage of an extremely high binding capacity must be carefully balanced with the quantity of antibody that the researcher is prepared to use to coat the agarose beads. Because antibodies can be a cost-limiting factor, it is best to calculate backward *from* the amount of protein that needs to be captured (depending upon the analysis to be performed downstream), *to* the amount of antibody that is required to bind that quantity of protein (with a small excess added in order to account for inefficiencies of the system), and back still further *to* the quantity of agarose that is needed to bind that particular quantity of antibody. In cases where antibody saturation is not required, this technology is unmatched in its ability to capture extremely large quantities of captured target proteins. The caveat here is that the *"high capacity advantage"* can become a *"high capacity disadvantage"* that is manifested when the enormous binding capacity of the sepharose/agarose beads is not completely saturated with antibodies. It often happens that the amount of antibody available to the researcher for their immunoprecipitation experiment is less than sufficient to saturate the agarose beads to be used in the immunoprecipitation. In these cases the researcher can end up with agarose particles that are only partially coated with antibodies, and the portion of the binding capacity of the agarose beads that is not coated with antibody is then free to bind anything that will stick, resulting in an elevated background signal due to non-specific binding of lysate components to the beads, which can make data interpretation difficult. While some may argue that for these reasons it is prudent to match the quantity of agarose (in terms of binding capacity) to the quantity of antibody that one wishes to be bound for the immunoprecipitation, a simple way to reduce the issue of non-specific binding to agarose beads and increase specificity is to preclear the lysate, which for any immunoprecipitation is highly recommended.

Preclearing

Lysates are complex mixtures of proteins, lipids, carbohydrates and nucleic acids, and one must assume that some amount of non-specific binding to the IP antibody, Protein A/G or the beaded support will occur and negatively affect the detection of the immunoprecipitated target(s). In most cases, *preclearing* the lysate at the start of each immunoprecipitation experiment is a way to remove potentially reactive components from the cell lysate prior to the immunoprecipitation to prevent the non-specific binding of these components to the IP beads or antibody. The basic preclearing procedure is described below, wherein the lysate is incubated with beads alone, which are then removed and discarded prior to the immunoprecipi-tation This approach, though, does not account for non-specific binding to the IP antibody, which can be considerable. Therefore, an alternative method of preclearing is to incubate the protein mixture with exactly the same components that will be used in the immunoprecipitation, except that a non-target, irrelevant antibody of the same antibody subclass as the IP antibody is used instead of the IP antibody itself. This approach attempts to use as close to the exact IP conditions and components as the actual immunoprecipitation to remove any non-specific cell constituent without capturing the target protein (unless, of course, the target protein non-specifically binds to some other IP component, which should be properly controlled for by analyzing the discarded beads used to preclear the lysate). The target protein can then be immunoprecipitated with the reduced risk of non-specific binding interfering with data interpretation.

Superparamagnetic Beads

While the vast majority of immunoprecipitations are performed with agarose beads, the use of superparamagnetic beads for immunoprecipitation is a much newer approach that is only recently gaining in popularity as an alternative to agarose beads for IP applications. Unlike agarose, magnetic beads are solid and can be spherical, depending on the type of bead, and antibody binding is limited to the surface of each bead. While these beads do not have the advantage of a porous center to increase the binding capacity, magnetic beads are significantly smaller than agarose beads (1 to 4μm), and the greater number of magnetic beads per volume than agarose beads collectively gives magnetic beads an effective surface area-to-volume ratio for optimum antibody binding.

Commercially available magnetic beads can be separated based by size uniformity into monodisperse and polydisperse beads. Monodisperse beads, also called microbeads, exhibit exact uniformity, and therefore all beads exhibit identical physical characteristics, including the binding capacity and the level of attraction to magnets. Polydisperse beads, while similar in size to monodisperse beads, show a wide range in size variability (1 to 4μm) that can influence their binding capacity and magnetic capture. Although both types of beads are commercially available for immunoprecipitation applications, the higher quality monodisperse superparamagnetic beads are more ideal for automatic protocols because of their consistent size, shape and performance. Monodisperse and polydisperse superparamagnetic beads are offered by many companies, including Invitrogen, Thermo Scientific, and Millipore.

Agarose Vs. Magnetic Beads

Proponents of magnetic beads claim that the beads exhibit a faster rate of protein binding over agarose beads for immunoprecipitation applications, although standard agarose bead-based immunoprecipitations have been performed in 1 hour. Claims have also been made that magnetic beads are better for immunoprecipitating extremely large protein complexes because of the complete lack of an upper size limit for such complexes, although there is no unbiased evidence stating this claim. The nature of magnetic bead technology does result in less sample handling due to the reduced physical stress on samples of magnetic separation versus repeated centrifugation when using agarose, which may contribute greatly to increasing the yield of labile (fragile) protein complexes. Additional factors, though, such as the binding capacity, cost of the reagent, the requirement of extra equipment and the capability to automate IP processes should be considered in the selection of an immunoprecipitation support.

Binding Capacity

Proponents of both agarose and magnetic beads can argue whether the vast difference in the binding capacities of the two beads favors one particular type of bead. In a bead-to-bead comparison, agarose beads have significantly greater surface area and therefore a greater binding capacity than magnetic beads due to the large bead size and sponge-like structure. But the variable pore size of the agarose causes a potential upper size limit that may affect the binding of extremely large proteins or protein complexes to internal binding sites, and therefore magnetic beads may be better suited for immunoprecipitating large proteins or protein complexes than agarose beads, although there is a lack of independent comparative evidence that proves either case.

Some argue that the significantly greater binding capacity of agarose beads may be a disadvantage because of the larger capacity of non-specific binding. Others may argue for the use of magnetic beads because of the greater quantity of antibody required to saturate the total binding capacity of agarose beads, which would obviously be an economical disadvantage of using agarose. While these arguments are correct outside the context of their practical use, these lines of reasoning ignore two key aspects of the principle of immunoprecipitation that demonstrates that the decision to use agarose or magnetic beads is not simply determined by binding capacity.

First, non-specific binding is not limited to the antibody-binding sites on the immobilized support; any surface of the antibody or component of the immunoprecipitation reaction can bind to nonspecific lysate constituents, and therefore nonspecific binding will still occur even when completely saturated beads are used. This is why it is important to preclear the sample before the immunoprecipitation is performed.

Second, the ability to capture the target protein is directly dependent upon the amount of immobilized antibody used, and therefore, in a side-by-side comparison of agarose and magnetic bead immunoprecipitation, the most protein that either support can capture is limited by the amount of antibody added. So the decision to saturate any type of support depends on the amount of protein required, as described above in the Agarose section of this page.

Cost

The price of using either type of support is a key determining factor in using agarose or magnetic beads for immunoprecipitation applications. A typical first-glance calculation on the cost of magnetic beads compared to sepharose beads may make the sepharose beads appear less expensive. But magnetic beads may be competitively priced compared to agarose for analytical-scale immunoprecipitations depending on the IP method used and the volume of beads required per IP reaction.

Using the traditional batch method of immunoprecipitation as listed below, where all components are added to a tube during the IP reaction, the physical handling characteristics of agarose beads necessitate a minimum quantity of beads for each IP experiment (typically in the range of 25 to 50μl beads per IP). This is because sepharose beads must be concentrated at the bottom of the tube by centrifugation and the supernatant removed after each incubation, wash, etc. This imposes absolute physical limitations on the process, as pellets of agarose beads less than 25 to 50μl are difficult if not impossible to visually identify at the bottom of the tube. With magnetic beads, there is no minimum quantity of beads required due to magnetic handling, and therefore, depending on the target antigen and IP antibody, it is possible to use considerably less magnetic beads.

Conversely, spin columns may be employed instead of normal microfuge tubes to significantly reduce the amount of agarose beads required per reaction. Spin columns contain a filter that allows all IP components except the beads to flow through using a brief centrifugation and therefore provide a method to use significantly less agarose beads with minimal loss.

Equipment

As mentioned above, only standard laboratory equipment is required for the use of agarose beads in immunoprecipitation applications, while high-power magnets are required for magnetic bead-based IP reactions. While the magnetic capture equipment may be cost-prohibitive, the rapid completion of immunoprecipitations using magnetic beads may be a financially beneficial approach when grants are due, because a 30-minute protocol with magnetic beads compared to overnight incubation at 4 °C with agarose beads may result in more data generated in a shorter length of time.

Automation

An added benefit of using magnetic beads is that automated immunoprecipitation devices are becoming more readily available. These devices not only reduce the amount of work and time to perform an IP, but they can also be used for high-throughput applications.

Summary

While clear benefits of using magnetic beads include the increased reaction speed, more gentle sample handling and the potential for automation, the choice of using agarose or magnetic beads based on the binding capacity of the support medium and the cost of the product may depend on the protein of interest and the IP method used. As with all assays, empirical testing is required to determine which method is optimal for a given application.

Protocol

Background

Once the solid substrate bead technology has been chosen, antibodies are coupled to the beads and the antibody-coated-beads can be added to the heterogeneous protein sample (e.g. homogenized tissue). At this point, antibodies that are immobilized to the beads will bind to the proteins that they specifically recognize. Once this has occurred the immunoprecipitation portion of the protocol is actually complete, as the specific proteins of interest are bound to the antibodies that are themselves immobilized to the beads. Separation of the immunocomplexes from the lysate is an extremely important series of steps, because the protein(s) must remain bound to each other (in the case of co-IP) and bound to the antibody during the wash steps to remove non-bound proteins and reduce background.

When working with agarose beads, the beads must be pelleted out of the sample by briefly spinning in a centrifuge with forces between 600–3,000 x g (times the standard gravitational force). This step may be performed in a standard microcentrifuge tube, but for faster separation, greater consistency and higher recoveries, the process is often performed in small spin columns with a pore size that allows liquid, but not agarose beads, to pass through. After centrifugation, the agarose beads will form a very loose fluffy pellet at the bottom of the tube. The supernatant containing contaminants can be carefully removed so as not to disturb the beads. The wash buffer can then be added to the beads and after mixing, the beads are again separated by centrifugation.

With superparamagnetic beads, the sample is placed in a magnetic field so that the beads can collect on the side of the tube. This procedure is generally complete in approximately 30 seconds, and the remaining (unwanted) liquid is pipetted away. Washes are accomplished by resuspending the beads (off the magnet) with the washing solution and then concentrating the beads back on the tube wall (by placing the tube back on the magnet). The washing is generally repeated several times to ensure adequate removal of contaminants. If the superparamagnetic beads are homogeneous in size and the magnet has been designed properly, the beads will concentrate uniformly on the side of the tube and the washing solution can be easily and completely removed.

After washing, the precipitated protein(s) are eluted and analyzed by gel electrophoresis, mass spectrometry, western blotting, or any number of other methods for identifying constituents in the complex. Protocol times for immunoprecipitation vary greatly due to a variety of factors, with protocol times increasing with the number of washes necessary or with the slower reaction kinetics of porous agarose beads.

Steps

1. Lyse cells and prepare sample for immunoprecipitation.
2. Pre-clear the sample by passing the sample over beads alone or bound to an irrelevant antibody to soak up any proteins that non-specifically bind to the IP components.
3. Incubate solution with antibody against the protein of interest. Antibody can be attached to solid support before this step (direct method) or after this step (indirect method). Continue the incubation to allow antibody-antigen complexes to form.
4. Precipitate the complex of interest, removing it from bulk solution.
5. Wash precipitated complex several times. Spin each time between washes when using agarose beads or place tube on magnet when using superparamagnetic beads and then remove the supernatant. After the final wash, remove as much supernatant as possible.
6. Elute proteins from the solid support using low-pH or SDS sample loading buffer.
7. Analyze complexes or antigens of interest. This can be done in a variety of ways:
    1. SDS-PAGE (sodium dodecyl sulfate-polyacrylamide gel electrophoresis) followed by gel staining.
    2. SDS-PAGE followed by: gel staining, cutting out individual stained protein bands, and sequencing the proteins in the bands by MALDI-Mass Spectrometry
    3. Transfer and Western Blot using another antibody for proteins that were interacting with the antigen followed by chemiluminesent visualization.

Protein Domain

A protein domain is a conserved part of a given protein sequence and (tertiary) structure that can evolve, function, and exist independently of the rest of the protein chain. Each domain forms a compact three-dimensional structure and often can be independently stable and folded. Many proteins consist of several structural domains. One domain may appear in a variety of different proteins. Molecular evolution uses domains as building blocks and these may be recombined in different arrangements to create proteins with different functions. Domains vary in length from between about 25 amino acids up to 500 amino acids in length. The shortest domains such as zinc fingers are stabilized by metal ions or disulfide bridges. Domains often form functional units, such as the calcium-binding EF hand domain of calmodulin. Because they are independently stable, domains can be "swapped" by genetic engineering between one protein and another to make chimeric proteins.

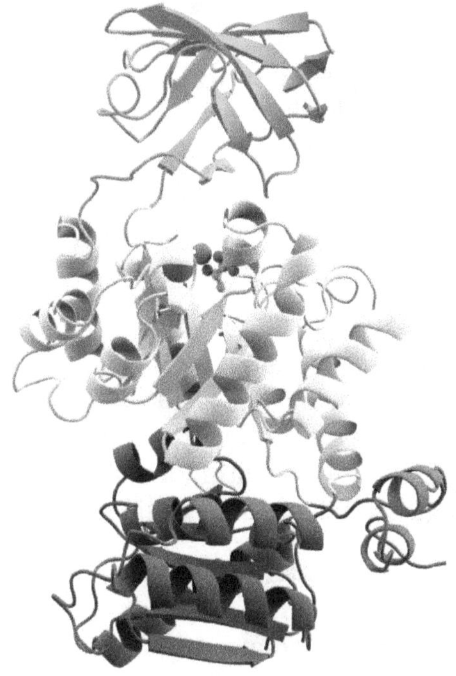

Pyruvate kinase, a protein with three domains (PDB: 1PKN).

Background

The concept of the domain was first proposed in 1973 by Wetlaufer after X-ray crystallographic studies of hen lysozyme and papain and by limited proteolysis studies of immunoglobulins. Wetlaufer defined domains as stable units of protein structure that could fold autonomously. In the past domains have been described as units of:

- compact structure
- function and evolution
- folding.

Each definition is valid and will often overlap, i.e. a compact structural domain that is found amongst diverse proteins is likely to fold independently within its structural environment. Nature often brings several domains together to form multidomain and multifunctional proteins with a vast number of possibilities. In a multidomain protein, each domain may fulfill its own function independently, or in a concerted manner with its neighbours. Domains can either serve as modules for building up large assemblies such as virus particles or muscle fibres, or can provide specific catalytic or binding sites as found in enzymes or regulatory proteins.

An appropriate example is pyruvate kinase, a glycolytic enzyme that plays an im-portant role in regulating the flux from fructose-1,6-biphosphate to pyruvate. It contains an all-β nucleotide binding domain (in blue), an α/β-substrate binding domain (in grey) and an α/β-regulatory domain (in olive green), connected by several polypeptide linkers. Each domain in this protein occurs in diverse sets of protein families.

The central α/β-barrel substrate binding domain is one of the most common enzyme folds. It is seen in many different enzyme families catalysing completely unrelated reactions. The α/β-barrel is commonly called the TIM barrel named after triose phosphate isomerase, which was the first such structure to be solved. It is currently classified into 26 homologous families in the CATH domain database. The TIM barrel is formed from a sequence of β-α-β motifs closed by the first and last strand hydrogen bonding together, forming an eight stranded barrel. There is debate about the evolutionary origin of this domain. One study has suggested that a single ancestral enzyme could have diverged into several families, while another suggests that a stable TIM-barrel structure has evolved through convergent evolution.

The TIM-barrel in pyruvate kinase is 'discontinuous', meaning that more than one segment of the polypeptide is required to form the domain. This is likely to be the result of the insertion of one domain into another during the protein's evolution. It has been shown from known structures that about a quarter of structural domains are discontinuous. The inserted β-barrel regulatory domain is 'continuous', made up of a single stretch of polypeptide.

Covalent association of two domains represents a functional and structural advantage since there is an increase in stability when compared with the same structures non-covalently associated. Other, advantages are the protection of intermediates within inter-domain enzymatic clefts that may otherwise be unstable in aqueous environments, and a fixed stoichiometric ratio of the enzymatic activity necessary for a sequential set of reactions.

Units of Protein Structure

The primary structure (string of amino acids) of a protein ultimately encodes its uniquely folded three-dimensional (3D) conformation. The most important factor governing the folding of a protein into 3D structure is the distribution of polar and non-polar side chains. Folding is driven by the burial of hydrophobic side chains into the interior of the molecule so to avoid contact with the aqueous environment. Generally proteins have a core of hydrophobic residues surrounded by a shell of hydrophilic residues. Since the peptide bonds themselves are polar they are neutralised by hydrogen bonding with each other when in the hydrophobic environment. This gives rise to regions of the polypeptide that form regular 3D structural patterns called secondary structure. There are two main types of secondary structure: α-helices and β-sheets.

Some simple combinations of secondary structure elements have been found to frequently occur in protein structure and are referred to as supersecondary structure or motifs. For example, the β-hairpin motif consists of two adjacent antiparallel β-strands joined by a small loop. It is present in most antiparallel β structures both as an isolated ribbon and as part of more complex β-sheets. Another common super-secondary structure is the β-α-β motif, which is frequently used to connect two parallel β-strands. The central α-helix connects the C-termini of the first strand to the N-termini of the second strand, packing its side chains against the β-sheet and therefore shielding the hydrophobic residues of the β-strands from the surface.

Structural alignment is an important tool for determining domains.

Tertiary Structure

Several motifs pack together to form compact, local, semi-independent units called domains. The overall 3D structure of the polypeptide chain is referred to as the protein's tertiary structure. Domains are the fundamental units of tertiary structure, each domain containing an individual hydrophobic core built from secondary structural units connected by loop regions. The packing of the polypeptide is usually much tighter in the interior than the exterior of the domain producing a solid-like core and a fluid-like surface. Core residues are often conserved in a protein family, whereas the residues in loops are less conserved, unless they are involved in the protein's function. Protein tertiary structure can be divided into four main classes based on the secondary structural content of the domain.

- All-α domains have a domain core built exclusively from α-helices. This class is dominated by small folds, many of which form a simple bundle with helices running up and down.
- All-β domains have a core composed of antiparallel β-sheets, usually two sheets packed against each other. Various patterns can be identified in the arrangement of the strands, often giving rise to the identification of recurring motifs, for example the Greek key motif.
- α+β domains are a mixture of all-α and all-β motifs. Classification of proteins into this class is difficult because of overlaps to the other three classes and therefore is not used in the CATH domain database.
- α/β domains are made from a combination of β-α-β motifs that predominantly form a parallel β-sheet surrounded by amphipathic α-helices. The secondary structures are arranged in layers or barrels.

Limits on Size

Domains have limits on size. The size of individual structural domains varies from 36 residues in E-selectin to 692 residues in lipoxygenase-1, but the majority, 90%, have less than 200 residues with an average of approximately 100 residues. Very short domains, less than 40 residues, are often stabilised by metal ions or disulfide bonds. Larger domains, greater than 300 residues, are likely to consist of multiple hydrophobic cores.

Quaternary Structure

Many proteins have a quaternary structure, which consists of several polypeptide chains that associate into an oligomeric molecule. Each polypeptide chain in such a protein is called a subunit. Hemoglobin, for example, consists of two α and two β subunits. Each of the four chains has an all-α globin fold with a heme pocket.

Domain swapping is a mechanism for forming oligomeric assemblies. In domain swapping, a secondary or tertiary element of a monomeric protein is replaced by the same element of another protein. Domain swapping can range from secondary structure elements to whole structural domains. It also represents a model of evolution for functional adaptation by oligomerisation, e.g. oligomeric enzymes that have their active site at subunit interfaces.

Domains as Evolutionary Modules

Nature is a tinkerer and not an inventor, new sequences are adapted from pre-existing sequences rather than invented. Domains are the common material used by nature to generate new sequences; they can be thought of as genetically mobile units, referred to as 'modules'. Often, the C and N termini of domains are close together in space, allowing them to easily be "slotted into" parent structures during the process of evolution. Many domain families are found in all three forms of life, Archaea, Bacteria and Eukarya. Domains that are repeatedly found in diverse proteins are often referred to as modules; examples can be found among extracellular proteins associated with clotting, fibrinolysis, complement, the extracellular matrix, cell surface adhesion molecules and cytokine receptors.

Molecular evolution gives rise to families of related proteins with similar sequence and structure. However, sequence similarities can be extremely low between proteins that share the same structure. Protein structures may be similar because proteins have diverged from a common ancestor. Alternatively, some folds may be more favored than others as they represent stable arrangements of secondary structures and some proteins may converge towards these folds over the course of evolution. There are currently about 110,000 experimentally determined protein 3D structures deposited within the Protein Data Bank (PDB). However, this set contains many identical or very similar structures. All proteins should be classified to structural families to understand their evolutionary relationships. Structural comparisons are best achieved at the domain level. For this reason many algorithms have been developed to automatically assign domains in proteins with known 3D structure.

The CATH domain database classifies domains into approximately 800 fold families; ten of these folds are highly populated and are referred to as 'super-folds'. Super-folds are defined as folds for which there are at least three structures without significant sequence similarity. The most populated is the α/β-barrel super-fold, as described previously.

Multidomain Proteins

The majority of genomic proteins, two-thirds in unicellular organisms and more than 80% in metazoa, are multidomain proteins created as a result of gene duplication events. Many domains in multidomain structures could have once existed as independent proteins. More and more domains in eukaryotic multidomain proteins can be found as independent proteins in prokaryotes. For example, vertebrates have a multi-enzyme polypeptide containing the GAR synthetase, AIR synthetase and GAR transformylase modules (GARs-AIRs-GARt; GAR: glycinamide ribonucleotide synthetase/transferase; AIR: aminoimidazole ribonucleotide synthetase). In insects, the polypeptide appears as GARs-(AIRs)2-GARt, in yeast GARs-AIRs is encoded separately from GARt, and in bacteria each domain is encoded separately.

Origin

Multidomain proteins are likely to have emerged from selective pressure during evolution to create new functions. Various proteins have diverged from common ancestors by different combinations and associations of domains. Modular units frequently move about, within and between biological systems through mechanisms of genetic shuffling:

- transposition of mobile elements including horizontal transfers (between species);
- gross rearrangements such as inversions, translocations, deletions and duplications;
- homologous recombination;
- slippage of DNA polymerase during replication.

Types of Organization

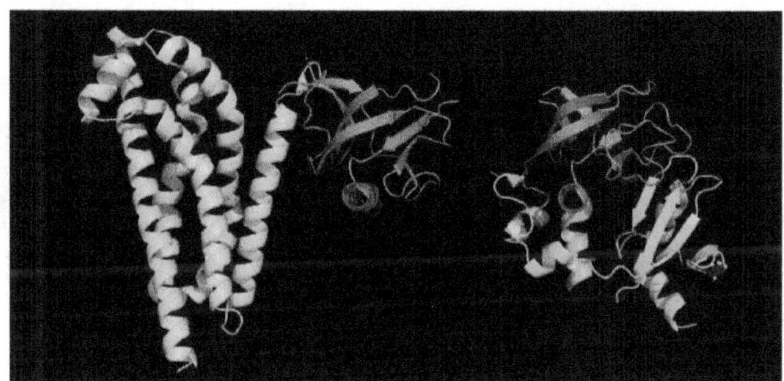

Insertions of similar PH domain modules (maroon) into two different proteins.

The simplest multidomain organization seen in proteins is that of a single domain repeated in tandem. The domains may interact with each other (domain-domain interaction) or remain isolated, like beads on string. The giant 30,000 residue muscle protein titin comprises about 120 fibronectin-III-type and Ig-type domains. In the serine proteases, a gene duplication event has led to the formation of a two β-barrel domain enzyme. The repeats have diverged so widely that there is no obvious sequence similarity between them. The active site is located at a cleft between the two β-barrel domains, in which functionally important residues are contributed from each domain. Genetically engineered mutants of the chymotrypsin serine protease were shown to have some proteinase activity even though their active site residues were abolished and it has therefore been postulated that the duplication event enhanced the enzyme's activity.

Modules frequently display different connectivity relationships, as illustrated by the kinesins and ABC transporters. The kinesin motor domain can be at either end of a polypeptide chain that includes a coiled-coil region and a cargo domain. ABC transporters are built with up to four domains consisting of two unrelated modules, ATP-binding cassette and an integral membrane module, arranged in various combinations.

Not only do domains recombine, but there are many examples of a domain having been inserted into another. Sequence or structural similarities to other domains demonstrate that homologues of inserted and parent domains can exist independently. An example is that of the 'fingers' inserted into the 'palm' domain within the polymerases of the Pol I family. Since a domain can be inserted into another, there should always be at least one continuous domain in a multidomain protein. This is the main difference between definitions of structural domains and evolutionary/functional domains. An evolutionary domain will be limited to one or two connections between domains, whereas structural domains can have unlimited connections, within a given criterion of the existence of a common core. Several structural domains could be assigned to an evolutionary domain.

A superdomain consists of two or more conserved domains of nominally independent origin, but subsequently inherited as a single structural/functional unit. This combined superdomain can occur in diverse proteins that are not related by gene duplication alone. An example of a superdomain is the protein tyrosine phosphatase–C2 domain pair in PTEN, tensin, auxilin and the membrane protein TPTE2. This superdomain is found in proteins in animals, plants and fungi. A key feature of the PTP-C2 superdomain is amino acid residue conservation in the domain interface.

Domains are Autonomous Folding Units

Folding

Protein folding - the unsolved problem : Since the seminal work of Anfinsen in the early 1960s, the goal to completely understand the mechanism by which a polypeptide rapidly folds into its stable native conformation remains elusive. Many experimental folding studies have contributed much to our understanding, but the principles that govern protein folding are still based on those discovered in the very first studies of folding. Anfinsen showed that the native state of a protein is thermodynamically stable, the conformation being at a global minimum of its free energy.

Folding is a directed search of conformational space allowing the protein to fold on a biologically feasible time scale. The Levinthal paradox states that if an averaged sized protein would sample all possible conformations before finding the one with the lowest energy, the whole process would take billions of years. Proteins typically fold within 0.1 and 1000 seconds. Therefore, the protein folding process must be directed some way through a specific folding pathway. The forces that direct this search are likely to be a combination of local and global influences whose effects are felt at various stages of the reaction.

Advances in experimental and theoretical studies have shown that folding can be viewed in terms of energy landscapes, where folding kinetics is considered as a progressive organisation of an ensemble of partially folded structures through which a protein passes on its way to the folded structure. This has been described in terms of a folding funnel, in which an unfolded protein has a large number of conformational states available and there are fewer states available to the folded protein. A funnel implies that for protein folding there is a decrease in energy and loss of entropy with increasing tertiary structure formation. The local roughness of the funnel reflects kinetic traps, corresponding to the accumulation of misfolded intermediates. A folding chain progresses toward lower intra-chain free-energies by increasing its compactness. The chain's conformational options become increasingly narrowed ultimately toward one native structure.

Advantage of Domains In Protein Folding

The organisation of large proteins by structural domains represents an advantage for protein folding, with each domain being able to individually fold, accelerating the folding process and reducing a potentially large combination of residue interactions. Furthermore, given the observed random distribution of hydrophobic residues in proteins, domain formation appears to be the optimal solution for a large protein to bury its hydrophobic residues while keeping the hydrophilic residues at the surface.

However, the role of inter-domain interactions in protein folding and in energetics of stabilisation of the native structure, probably differs for each protein. In T4 lysozyme, the influence of one domain on the other is so strong that the entire molecule is resistant to proteolytic cleavage. In this case, folding is a sequential process where the C-terminal domain is required to fold independently in an early step, and the other domain requires the presence of the folded C-terminal domain for folding and stabilisation.

It has been found that the folding of an isolated domain can take place at the same rate or sometimes faster than that of the integrated domain, suggesting that unfavourable interactions with the rest of the protein can occur during folding. Several arguments suggest that the slowest step in the folding of large proteins is the pairing of the folded domains. This is either because the domains are not folded entirely correctly or because the small adjustments required for their interaction are energetically unfavourable, such as the removal of water from the domain interface.

Domains and Protein Flexibility

Protein domain dynamics play a key role in a multitude of molecular recognition and signaling processes. Protein domains, connected by intrinsically disordered flexible linker domains, induce long-range allostery via protein domain dynamics. The resultant dynamic modes cannot be generally predicted from static structures of either the entire protein or individual domains.

Domain Definition from Structural Co-ordinates

The importance of domains as structural building blocks and elements of evolution has brought about many automated methods for their identification and classification in proteins of known structure. Automatic procedures for reliable domain assignment is essential for the generation of the domain databases, especially as the number of known protein structures is increasing. Although the boundaries of a domain can be determined by visual inspection, construction of an automated method is not straightforward. Problems occur when faced with domains that are discontinuous or highly associated. The fact that there is no standard definition of what a domain really is has meant that domain assignments have varied enormously, with each researcher using a unique set of criteria.

A structural domain is a compact, globular sub-structure with more interactions within it than with the rest of the protein. Therefore, a structural domain can be determined by two visual characteristics: its compactness and its extent of isolation. Measures of local compactness in proteins have been used in many of the early methods of domain assignment and in several of the more recent methods.

Methods

One of the first algorithms used a Cα-Cα distance map together with a hierarchical clustering routine that considered proteins as several small segments, 10 residues in length. The initial segments were clustered one after another based on inter-segment distances; segments with the shortest distances were clustered and considered as single segments thereafter. The stepwise clustering finally included the full protein. Go also exploited the fact that inter-domain distances are normally larger than intra-domain distances; all possible Cα-Cα distances were represented as diagonal plots in which there were distinct patterns for helices, extended strands and combinations of secondary structures.

The method by Sowdhamini and Blundell clusters secondary structures in a protein based on their Cα-Cα distances and identifies domains from the pattern in their dendrograms. As the procedure does not consider the protein as a continuous chain of amino acids there are no problems in treating discontinuous domains. Specific nodes in these dendrograms are identified as tertiary structural clusters of the protein, these include both super-secondary structures and domains. The DOMAK algorithm is used to create the 3Dee domain database. It calculates a 'split value' from the number of each type of contact when the protein is divided arbitrarily into two parts. This split value is large when the two parts of the structure are distinct.

The method of Wodak and Janin was based on the calculated interface areas between two chain segments repeatedly cleaved at various residue positions. Interface areas were calculated by comparing surface areas of the cleaved segments with that of the native structure. Potential domain boundaries can be identified at a site where the interface area was at a minimum. Other methods have used measures of solvent accessibility to calculate compactness.

The PUU algorithm incorporates a harmonic model used to approximate inter-domain dynamics. The underlying physical concept is that many rigid interactions will occur within each domain and loose interactions will occur between domains. This algorithm is used to define domains in the FSSP domain database.

Swindells (1995) developed a method, DETECTIVE, for identification of domains in protein structures based on the idea that domains have a hydrophobic interior. Deficiencies were found to occur when hydrophobic cores from different domains continue through the interface region.

RigidFinder is a novel method for identification of protein rigid blocks (domains and loops) from two different conformations. Rigid blocks are defined as blocks where all inter residue distances are conserved across conformations.

A general method to identify *dynamical domains*, that is protein regions that behave approximately as rigid units in the course of structural fluctuations, has been introduced by Potestio et al. and, among other applications was also used to compare the consistency of the dynamics-based domain subdivisions with standard structure-based ones. The method, termed PiSQRD, is publicly available in the form of a webserver. The latter allows users to optimally subdivide single-chain or multimeric proteins into quasi-rigid domains based on the collective modes of fluctuation of the system. By default the latter are calculated through an elastic network model; alternatively pre-calculated essential dynamical spaces can be uploaded by the user.

Example Domains

- Armadillo repeats : named after the β-catenin-like Armadillo protein of the fruit fly *Drosophila*.

- Basic Leucine zipper domain (bZIP domain) : is found in many DNA-binding eukaryotic proteins. One part of the domain contains a region that mediates sequence-specific DNA-binding properties and the Leucine zipper that is required for the dimerization of two DNA-binding regions. The DNA-binding region comprises a number of basic aminoacids such as arginine and lysine
- Cadherin repeats : Cadherins function as Ca^{2+}-dependent cell-cell adhesion proteins. Cadherin domains are extracellular regions which mediate cell-to-cell homophilic binding between cadherins on the surface of adjacent cells.
- Death effector domain (DED) : allows protein-protein binding by homotypic interactions (DED-DED). Caspase proteases trigger apoptosis via proteolytic cascades. Pro-Caspase-8 and pro-caspase-9 bind to specific adaptor molecules via DED domains and this leads to autoactivation of caspases.
- EF hand : a helix-turn-helix structural motif found in each structural domain of the signaling protein calmodulin and in the muscle protein troponin-C.
- Immunoglobulin-like domains : are found in proteins of the immunoglobulin superfamily (IgSF). They contain about 70-110 amino acids and are classified into different categories (IgV, IgC1, IgC2 and IgI) according to their size and function. They possess a characteristic fold in which two beta sheets form a “sandwich” that is stabilized by interactions between conserved cysteines and other charged amino acids. They are important for protein-to-protein interactions in processes of cell adhesion, cell activation, and molecular recognition. These domains are commonly found in molecules with roles in the immune system.
- Phosphotyrosine-binding domain (PTB) : PTB domains usually bind to phosphorylated tyrosine residues. They are often found in signal transduction proteins. PTB-domain binding specificity is determined by residues to the amino-terminal side of the phosphotyrosine. Examples: the PTB domains of both SHC and IRS-1 bind to a NPXpY sequence. PTB-containing proteins such as SHC and IRS-1 are important for insulin responses of human cells.
- Pleckstrin homology domain (PH) : PH domains bind phosphoinositides with high affinity. Specificity for PtdIns(3)P, PtdIns(4)P, PtdIns(3,4)P2, PtdIns(4,5)P2, and PtdIns(3,4,5)P3 have all been observed. Given the fact that phosphoinositides are sequestered to various cell membranes (due to their long lipophilic tail) the PH domains usually causes recruitment of the protein in question to a membrane where the protein can exert a certain function in cell signalling, cytoskeletal reorganization or membrane trafficking.
- Src homology 2 domain (SH2) : SH2 domains are often found in signal transduction proteins. SH2 domains confer binding to phosphorylated tyrosine (pTyr). Named after the phosphotyrosine binding domain of the src viral oncogene, which is itself a tyrosine kinase.
- Zinc finger DNA binding domain (ZnF_GATA) : ZnF_GATA domain-containing proteins are typically transcription factors that usually bind to the DNA sequence [AT]GATA[AG] of promoters.

Domains of Unknown Function

A large fraction of domains are of unknown function. A domain of unknown function (DUF) is a protein domain that has no characterized function. These families have been collected together in the Pfam database using the prefix DUF followed by a number, with examples being DUF2992 and DUF1220. There are now over 3,000 DUF families within the Pfam database representing over 20% of known families.

Protein Biosynthesis

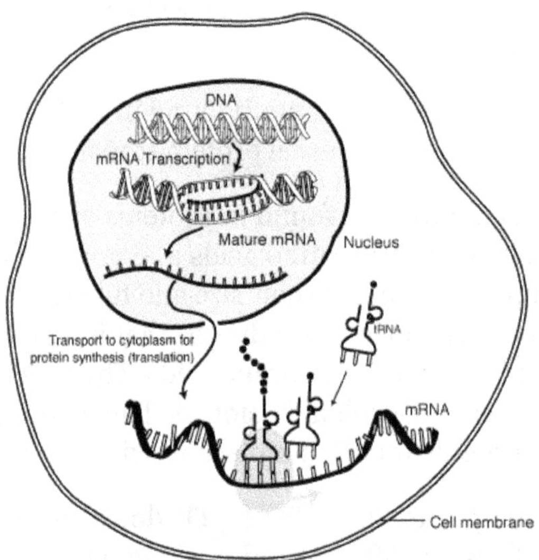

RNA is transcribed in the nucleus; once completely processed, it is transported to the cytoplasm and translated by the ribosome (shown in very pale grey behind the tRNA).

Protein biosynthesis is the process whereby biological cells generate new proteins; it is balanced by the loss of cellular proteins via degradation or export. Translation, the assembly of amino acids by ribosomes, is an essential part of the biosynthetic pathway, along with generation of messenger RNA (mRNA), aminoacylation of transfer RNA (tRNA), co-translational transport, and post-translational modification. Protein biosynthesis is strictly regulated at multiple steps. They are principally during transcription (phenomena of RNA synthesis from DNA template) and translation (phenomena of amino acid assembly from RNA).

The cistron DNA is transcribed into the first of a series of RNA intermediates. The last version is used as a template in synthesis of a polypeptide chain. Protein will often be synthesized directly from genes by translating mRNA. However, when a protein must be available on short notice or in large quantities, a protein precursor is produced. A proprotein is an inactive protein containing one or more inhibitory peptides that can be activated when the inhibitory sequence is removed by proteolysis during posttranslational modification. A preprotein is a form that contains a signal sequence (an N-terminal signal peptide) that specifies its insertion into or through membranes, i.e., targets them for secretion. The signal peptide is cleaved off in the endoplasmic reticulum. Preproproteins have both sequences (inhibitory and signal) still present.

In protein synthesis, a succession of tRNA molecules charged with appropriate amino acids are brought together with an mRNA molecule and matched up by base-pairing through the anti-codons of the tRNA with successive codons of the mRNA. The amino acids are then linked together to extend the growing protein chain, and the tRNAs, no longer carrying amino acids, are released. This whole complex of processes is carried out by the ribosome, formed of two main chains of RNA, called ribosomal RNA (rRNA), and more than 50 different proteins. The ribosome latches onto the end of an mRNA molecule and moves along it, capturing loaded tRNA molecules and joining together their amino acids to form a new protein chain.

Protein biosynthesis, although very similar, is different for prokaryotes and eukaryotes.

Transcription

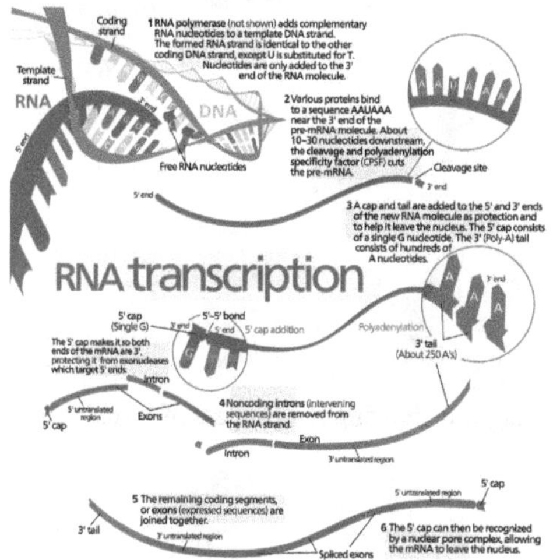

Diagram showing the process of transcription

In transcription an mRNA chain is generated, with one strand of the DNA double helix in the genome as a template. This strand is called the template strand. Transcription can be divided into 3 stages: initiation, elongation, and termination, each regulated by a large number of proteins such as transcription factors and coactivators that ensure that the correct gene is transcribed.

Transcription occurs in the cell nucleus, where the DNA is held. The DNA structure of the cell is made up of two helixes made up of sugar and phosphate held together by hydrogen bonds between the bases of opposite strands. The sugar and the phosphate in each strand are joined together by stronger phosphodiester covalent bonds. The DNA is "unzipped" (disruption of hydrogen bonds between different single strands) by the enzyme helicase, leaving the single nucleotide chain open to be copied. RNA polymerase reads the DNA strand from the 3-prime (3') end to the 5-prime (5') end, while it synthesizes a single strand of messenger RNA in the 5'-to-3' direction. The general RNA structure is very similar to the DNA structure, but in RNA the nucleotide uracil takes the place that thymine occupies in DNA. The single strand of mRNA leaves the nucleus through nuclear pores, and migrates into the cytoplasm.

The first product of transcription differs in prokaryotic cells from that of eukaryotic cells, as in prokaryotic cells the product is mRNA, which needs no post-transcriptional modification, whereas, in eukaryotic cells, the first product is called primary transcript, that needs post-transcriptional modification (capping with 7-methyl-guanosine, tailing with a poly A tail) to give hnRNA (heterogeneous nuclear RNA). hnRNA then undergoes splicing of introns (noncoding parts of the gene) via spliceosomes to produce the final mRNA.

Translation

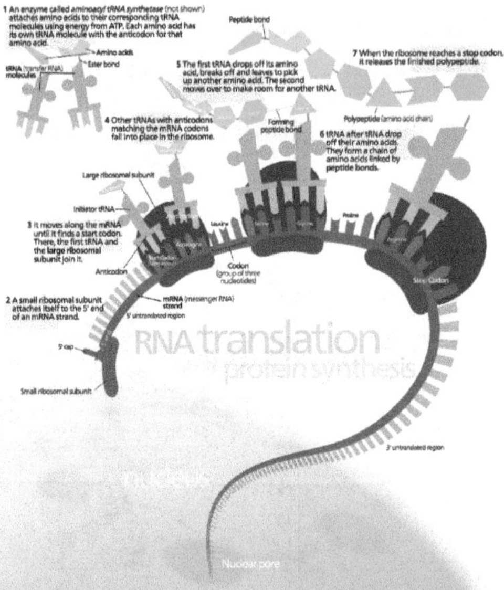

Diagram showing the process of translation

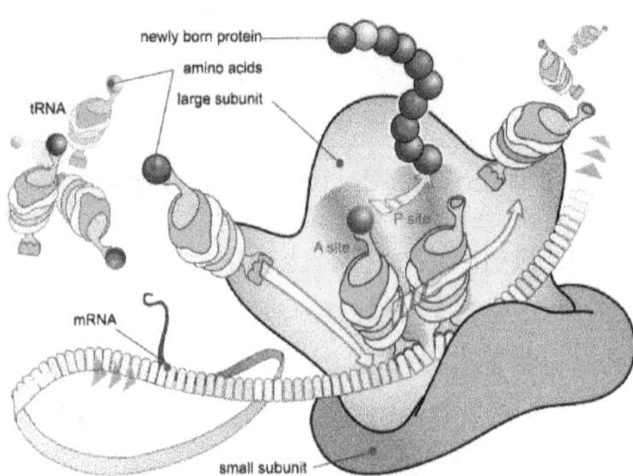

Diagram showing the translation of mRNA and the synthesis of proteins by a ribosome

The synthesis of proteins from RNA is known as translation. In eukaryotes, translation occurs in the cytoplasm, where the ribosomes are located. Ribosomes are made of a small and large subunit that surround the mRNA. In translation, messenger RNA (mRNA) is decoded to produce a specific polypeptide according to the rules specified by the trinucleotide genetic code. This uses an mRNA sequence as a template to guide the synthesis of a chain of amino acids that form a protein. Translation proceeds in four phases: activation, initiation, elongation, and termination (all describing the growth of the amino acid chain, or polypeptide that is the product of translation).

In activation, the correct amino acid (AA) is joined to the correct transfer RNA (tRNA). While this is not, in the technical sense, a step in translation, it is required for translation to proceed. The AA is joined by its carboxyl group to the 3' OH of the tRNA by an ester bond. When the tRNA has an amino acid linked to it, it is termed "charged". Initiation involves the small subunit of the ribosome binding to 5' end of mRNA with the help of initiation factors (IF), other proteins that assist the process. Elongation occurs when the next aminoacyl-tRNA (charged tRNA) in line binds to the ribosome along with GTP and an elongation factor. Termination of the polypeptide happens when the A site of the ribosome faces a stop codon (UAA, UAG, or UGA). When this happens, no tRNA can recognize it, but releasing factor can recognize nonsense codons and causes the release of the polypeptide chain. The capacity of disabling or inhibiting translation in protein biosynthesis is used by some antibiotics such as anisomycin, cycloheximide, chloramphenicol, tetracycline, streptomycin, erythromycin, puromycin, etc.

Events Following Protein Translation

The events following biosynthesis include post-translational modification and protein folding. During and after synthesis, polypeptide chains often fold to assume, so called, native secondary and tertiary structures. This is known as *protein folding.*

Protein Structure

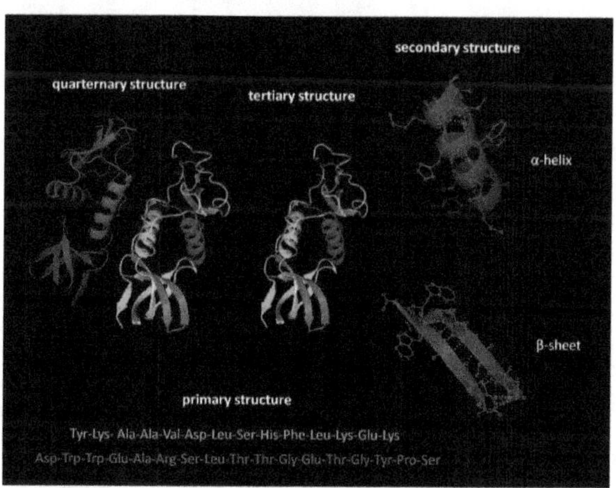

Protein structure is the three-dimensional arrangement of atoms in a protein molecule. Proteins are polymers — specifically polypeptides — formed from sequences of amino acids, the monomers of the polymer. A single amino acid monomer may also be called a residue (chemistry) indicating a repeating unit of a polymer. Proteins form by amino acids undergoing condensation reactions, in which the amino acids lose one water molecule per reaction in order to attach to one another with a peptide bond. By convention, a chain under 30 amino acids is often identified as a peptide, rather than a protein. To be able to perform their biological function, proteins fold into one or more specific spatial conformations driven by a number of non-covalent interactions such as hydrogen bonding, ionic interactions, Van der Waals forces, and hydrophobic packing. To understand the functions of proteins at a molecular level, it is often necessary to determine their three-dimensional structure. This is the topic of the scientific field of structural biology, which employs techniques such as X-ray crystallography, NMR spectroscopy, and dual polarisation interferometry to determine the structure of proteins.

Protein structures range in size from tens to several thousand amino acids. By physical size, proteins are classified as nanoparticles, between 1–100 nm. Very large aggregates can be formed from protein subunits. For example, many thousands of actin molecules assemble into a microfilament.

A protein may undergo reversible structural changes in performing its biological function. The alternative structures of the same protein are referred to as different conformational isomers, or simply, conformations, and transitions between them are called conformational changes.

Levels of Protein Structure

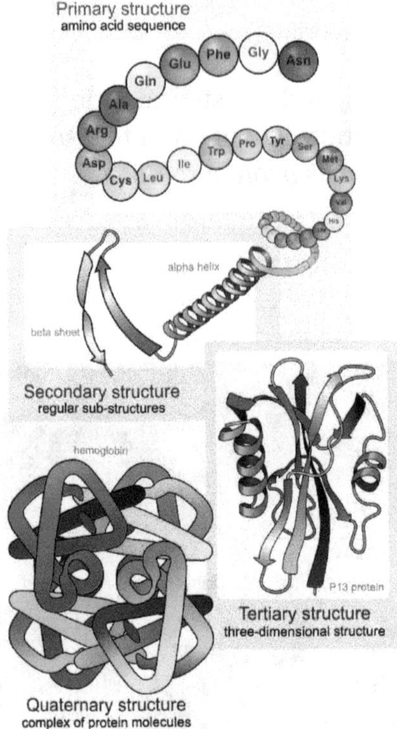

Protein structure, from primary to quaternary structure.

There are four distinct levels of protein structure.

Amino Acid Residues

Each α-amino acid consists of a backbone that is present in all the amino acid types and a side chain that is unique to each type of residue. An exception from this rule is proline. Because the carbon atom is bound to four different groups it is chiral, however only one of the isomers occur in biological proteins. Glycine however, is not chiral since its side chain is a hydrogen atom. A simple mnemonic for correct L-form is "CORN": when the C_α atom is viewed with the H in front, the residues read "CO-R-N" in a clockwise direction.

Primary Structure

The primary structure of a protein refers to the linear sequence of amino acids in the polypeptide chain. The primary structure is held together by covalent bonds such as peptide bonds, which are made during the process of protein biosynthesis. The two ends of the polypeptide chain are referred to as the carboxyl terminus (C-terminus) and the amino terminus (N-terminus) based on the nature of the free group on each extremity. Counting of residues always starts at the N-terminal end (NH_2-group), which is the end where the amino group is not involved in a peptide bond. The primary structure of a protein is determined by the gene corresponding to the protein. A specific sequence of nucleotides in DNA is transcribed into mRNA, which is read by the ribosome in a process called translation. The sequence of amino acids in insulin was discovered by Frederick Sanger, establishing that proteins have defining amino acid sequences. The sequence of a protein is unique to that protein, and defines the structure and function of the protein. The sequence of a protein can be determined by methods such as Edman degradation or tandem mass spectrometry. Often, however, it is read directly from the sequence of the gene using the genetic code. It is strictly recommended to use the words "amino acid residues" when discussing proteins because when a peptide bond is formed, a water molecule is lost, and therefore proteins are made up of amino acid residues. Post-translational modification such as disulfide bond formation, phosphorylations and glycosylations are usually also considered a part of the primary structure, and cannot be read from the gene. For example, insulin is composed of 51 amino acids in 2 chains. One chain has 31 amino acids, and the other has 20 amino acids.

Secondary Structure

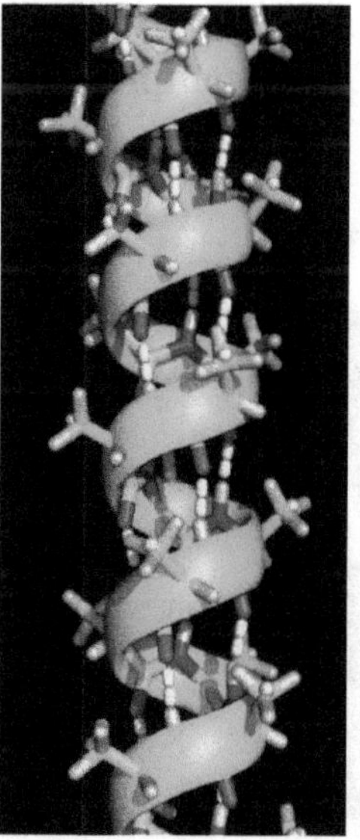

An α-helix with hydrogen bonds (yellow dots)

Secondary structure refers to highly regular local sub-structures on the actual polypeptide backbone chain. Two main types of secondary structure, the α-helix and the β-strand or β-sheets, were suggested in 1951 by Linus Pauling and coworkers. These secondary structures are defined by patterns of hydrogen bonds between the main-chain peptide groups. They have a regular geometry, being constrained to specific values of the dihedral angles ψ and φ on the Ramachandran plot. Both the α-helix and the β-sheet represent a way of saturating all the hydrogen bond donors and acceptors in the peptide backbone. Some parts of the protein are ordered but do not form any regular structures. They should not be confused with random coil, an unfolded polypeptide chain lacking any fixed three-dimensional structure. Several sequential secondary structures may form a "supersecondary unit".

Tertiary Structure

Tertiary structure refers to the three-dimensional structure of monomeric and multimeric protein molecules. The α-helixes and β-pleated-sheets are folded into a compact globular structure. The folding is driven by the *non-specific* hydrophobic interactions, the burial of hydrophobic residues from water, but the structure is stable only when the parts of a protein domain are locked into place by *specific* tertiary interactions, such as salt bridges, hydrogen bonds, and the tight packing of side chains and disulfide bonds. The disulfide bonds are extremely rare in cytosolic proteins, since the cytosol (intracellular fluid) is generally a reducing environment.

Quaternary Structure

Quaternary structure is the three-dimensional structure of a multi-subunit protein and how the subunits fit together. In this context, the quaternary structure is stabilized by the same non-covalent interactions and disulfide bonds as the tertiary structure. Complexes of two or more polypeptides (i.e. multiple subunits) are called multimers. Specifically it would be called a dimer if it contains two subunits, a trimer if it contains three subunits, a tetramer if it contains four subunits, and a pentamer if it contains five subunits. The subunits are frequently related to one another by symmetry operations, such as a 2-fold axis in a dimer. Multimers made up of identical subunits are referred to with a prefix of "homo-" (e.g. a homotetramer) and those made up of different subunits are referred to with a prefix of "hetero-", for example, a heterotetramer, such as the two alpha and two beta chains of hemoglobin.

Domains, Motifs, and Folds in Protein Structure

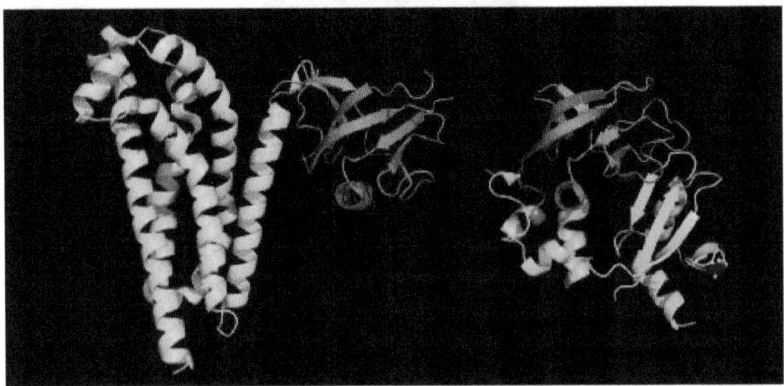

Protein domains. The two shown protein structures share a common domain (maroon), the PH domain, which is involved in phosphatidylinositol (3,4,5)-trisphosphate binding

Proteins are frequently described as consisting of several structural units. These units include domains, motifs, and folds. Despite the fact that there are about 100,000 different proteins expressed in eukaryotic systems, there are many fewer different domains, structural motifs and folds.

Structural Domain

A structural domain is an element of the protein's overall structure that is self-stabilizing and often folds independently of the rest of the protein chain. Many domains are not unique to the protein products of one gene or one gene family but instead appear in a variety of proteins. Domains often are named and singled out because they figure prominently in the biological function of the protein they belong to; for example, the "calcium-binding domain of calmodulin". Because they are independently stable, domains can be "swapped" by genetic engineering between one protein and another to make chimera proteins.

Structural and Sequence Motif

The structural and sequence motifs refer to short segments of protein three-dimensional structure or amino acid sequence that were found in a large number of different proteins.

Supersecondary Structure

The supersecondary structure refers to a specific combination of secondary structure elements, such as β-α-β units or a helix-turn-helix motif. Some of them may be also referred to as structural motifs.

Protein Fold

A protein fold refers to the general protein architecture, like a helix bundle, β-barrel, Rossman fold or different "folds" provided in the Structural Classification of Proteins database. A related concept is protein topology that refers to the arrangement of contacts within the protein.

Superdomain

A superdomain consists of two or more nominally unrelated structural domains that are inherited as a single unit and occur in different proteins. An example is provided by the protein tyrosine phosphatase domain and C2 domain pair in PTEN, several tensin proteins, auxilin and proteins in plants and fungi. The PTP-C2 superdomain evidently came into existence prior to the divergence of fungi, plants and animals is therefore likely to be about 1.5 billion years old.

Protein Folding

Once translated by a ribosome, each polypeptide folds into its characteristic three-dimensional structure from a random coil. Since the fold is maintained by a network of interactions between amino acids in the polypeptide, the native state of the protein chain is determined by the amino acid sequence (Anfinsen's dogma).

Protein Structure Determination

Around 90% of the protein structures available in the Protein Data Bank have been determined by X-ray crystallography. This method allows one to measure the three-dimensional (3-D) density distribution of electrons in the protein, in the crystallized state, and thereby infer the 3-D coordinates of all the atoms to be determined to a certain resolution. Roughly 9% of the known protein structures have been obtained by nuclear magnetic resonance techniques. The secondary structure composition can be determined via circular dichroism. Vibrational spectroscopy can also be used to characterize the conformation of peptides, polypeptides, and proteins. Two-dimensional infrared spectroscopy has become a valuable method to investigate the structures of flexible peptides and proteins that cannot be studied with other methods. Cryo-electron microscopy has recently become a means of determining protein structures to high resolution, less than 5 ångströms or 0.5 nanometer, and is anticipated to increase in power as a tool for high resolution work in the next decade. This technique is still a valuable resource for researchers working with very large protein complexes such as virus coat proteins and amyloid fibers. A more qualitative picture of protein structure is often obtained by proteolysis, which is also useful to screen for more crystallizable protein samples. Novel implementations of this approach, including fast parallel proteolysis (FASTpp), can probe the structured fraction and its stability without the need for purification.

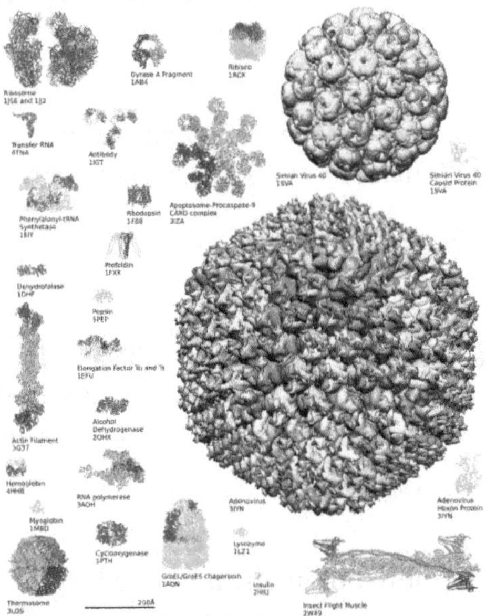

Examples of protein structures from the PDB

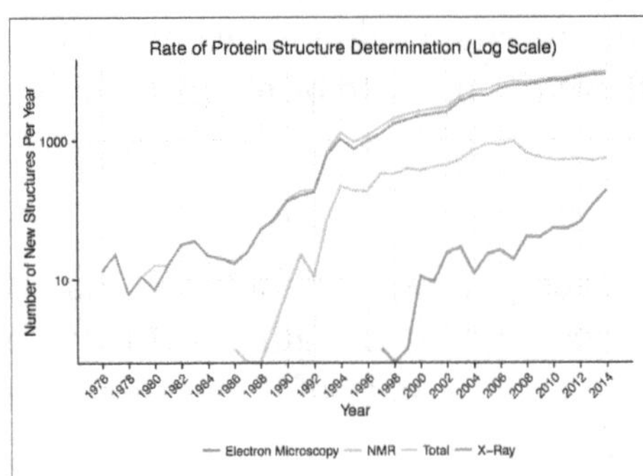

Rate of Protein Structure Determination by Method and Year

Protein Sequence Analysis: Ensembles

Proteins are often thought of as relatively stable structures that have a set tertiary structure and experience conformational changes as a result of being modified by other proteins or as part of enzymatic activity. However proteins have varying degrees of stability and some of the less stable variants are intrinsically disordered proteins. These proteins exist and function in a relatively 'disordered' state lacking a stable tertiary structure. As a result, they are difficult to describe in a standard protein structure model that was designed for proteins with a fixed tertiary structure. Conformational ensembles have been devised as a way to provide a more accurate and 'dynamic' representation of the conformational state of intrinsically disordered proteins. Conformational ensembles function by attempting to represent the various conformations of intrinsically disordered proteins within an ensemble file (the type found at the Protein Ensemble Database).

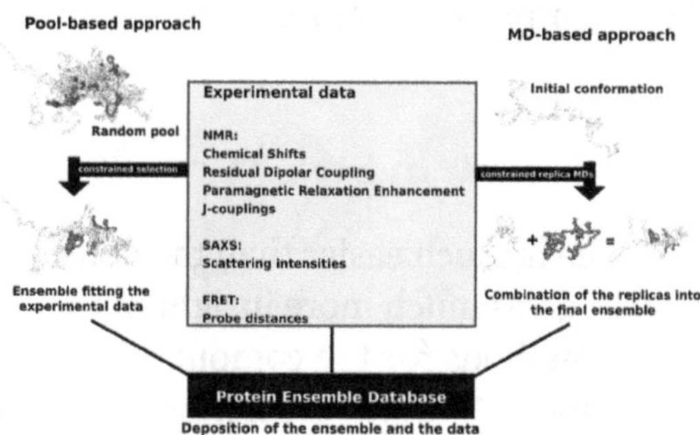

Schematic view of the two main ensemble modeling approaches.

Protein ensemble files are a representation of a protein that can be considered to have a flexible structure. Creating these files requires determining which of the various theoretically possible protein conformations actually exist. One approach is to apply computational algorithms to the protein data in order to try to determine the most likely set of conformations for an ensemble file.

There are multiple methods for preparing data for the Protein Ensemble Database that fall into two general methodologies – pool and molecular dynamics (MD) approaches (diagrammed in the figure). The pool based approach uses the protein's amino acid sequence to create a massive pool of random conformations. This pool is then subjected to more computational processing that creates a set of theoretical parameters for each conformation based on the structure. Conformational subsets from this pool whose average theoretical parameters closely match known experimental data for this protein are selected.

The molecular dynamics approach takes multiple random conformations at a time and subjects all of them to experimental data. Here the experimental data is serving as limitations to be placed on the conformations (e.g. known distances between atoms). Only conformations that manage to remain within the limits set by the experimental data are accepted. This approach often applies large amounts of experimental data to the conformations which is a very computationally demanding task.

Protein	Data Type	Protocol	PED ID	References
Sic1/Cdc4	NMR and SAXS	Pool-based	PED9AAA	
p15 PAF	NMR and SAXS	Pool-based	PED6AAA	
MKK7	NMR	Pool-based	PED5AAB	
Beta-synuclein	NMR	MD-based	PED1AAD	
P27 KID	NMR	MD-based	PED2AAA	

(adapted from image in "Computational approaches for inferring the functions of intrinsically disordered proteins")

Structure Classification

Protein structures can be grouped based on their similarity or a common evolutionary origin.

The Structural Classification of Proteins database and CATH database provide two different structural classifications of proteins. Shared structure between proteins is considered evidence of evolutionary relatedness between proteins and is used group proteins together into protein superfamilies.

Computational Prediction of Protein Structure

The generation of a protein sequence is much easier than the determination of a protein structure. However, the structure of a protein gives much more insight in the function of the protein than its sequence. Therefore, a number of methods for the computational prediction of protein structure from its sequence have been developed. *Ab initio* prediction methods use just the sequence of the protein. Threading and homology modeling methods can build a 3-D model for a protein of unknown structure from experimental structures of evolutionarily-related proteins, called a protein family.

Protein Structure Prediction

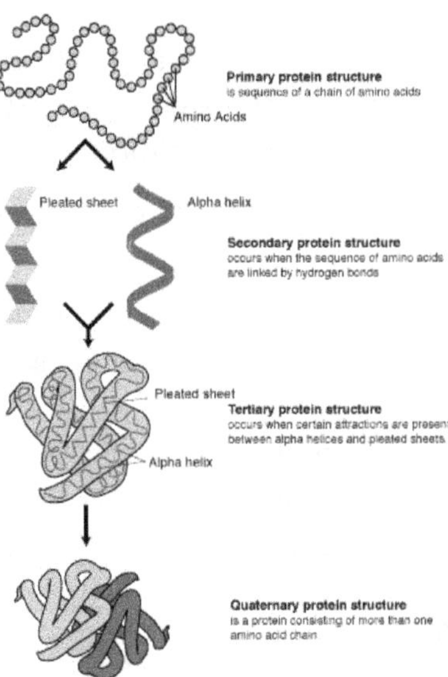

Constituent amino-acids can be analyzed to predict secondary, tertiary and quaternary protein structure.

Protein structure prediction is the inference of the three-dimensional structure of a protein from its amino acid sequence — that is, the prediction of its folding and its secondary and tertiary structure from its primary structure. Structure prediction is fundamentally different from the inverse problem of protein design. Protein structure prediction is one of the most important goals pursued by bioinformatics and theoretical chemistry; it is highly important in medicine (for example, in drug design) and biotechnology (for example, in the design of novel enzymes). Every two years, the performance of current methods is assessed in the CASP experiment (Critical Assessment of Techniques for Protein Structure Prediction). A continuous evaluation of protein structure prediction web servers is performed by the community project CAMEO3D.

Protein Structure and Terminology

Proteins are chains of amino acids joined together by peptide bonds. Many conformations of this chain are possible due to the rotation of the chain about each Cα atom. It is these conformational changes that are responsible for differences in the three dimensional structure of proteins. Each amino acid in the chain is polar, i.e. it has separated positive and negative charged regions with a free C=O group, which can act as hydrogen bond acceptor and an NH group, which can act as hydrogen bond donor. These groups can therefore interact in the protein structure. The 20 amino acids can be classified according to the chemistry of the side chain which also plays an important structural role. Glycine takes on a special position, as it has the smallest side chain, only one Hydrogen atom, and therefore can increase the local flexibility in the protein structure. Cysteine on the other hand can react with another cysteine residue and thereby form a cross link stabilizing the whole structure.

The protein structure can be considered as a sequence of secondary structure elements, such as α helices and β sheets, which together constitute the overall three-dimensional configuration of the protein chain. In these secondary structures regular patterns of H bonds are formed between neighboring amino acids, and the amino acids have similar Φ and Ψ angles.

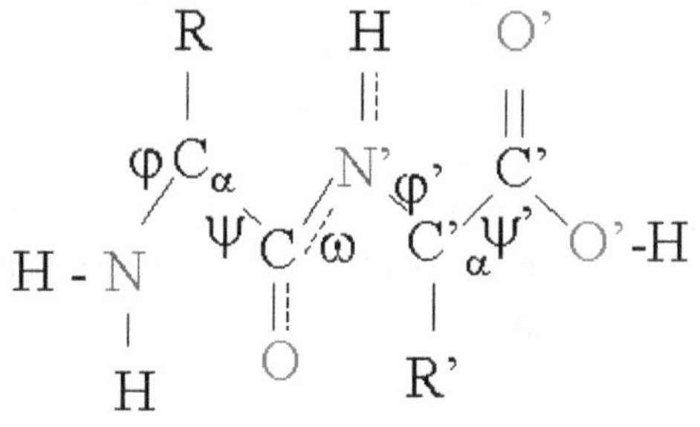

```
            R         H       O’
            |         ||      ||
          φ C_α      N’ φ’    C’
           /    ψ C ω   C’_α ψ’  O’-H
     H - N        ||     |
         |        O      R’
         H
```

Bond angles for ψ and ω

The formation of these structures neutralizes the polar groups on each amino acid. The secondary structures are tightly packed in the protein core in a hydrophobic environment. Each amino acid side group has a limited volume to occupy and a limited number of possible interactions with other nearby side chains, a situation that must be taken into account in molecular modeling and alignments.

α Helix

The α helix is the most abundant type of secondary structure in proteins. The α helix has 3.6 amino acids per turn with an H bond formed between every fourth residue; the average length is 10 amino acids (3 turns) or 10 Å but varies from 5 to 40 (1.5 to 11 turns). The alignment of the H bonds creates a dipole moment for the helix with a resulting partial positive charge at the amino end of the helix. Because this region has free NH2 groups, it will interact with negatively charged groups such as phosphates. The most common location of α helices is at the surface of protein cores, where they provide an interface with the aqueous environment. The inner-facing side of the helix tends to have hydrophobic amino acids and the outer-facing side hydrophilic amino acids. Thus, every third of four amino acids along the chain will tend to be hydrophobic, a pattern that can be quite readily detected. In the leucine zipper motif, a repeating pattern of leucines on the facing sides of two adjacent helices is highly predictive of the motif. A helical-wheel plot can be used to show this repeated pattern. Other α helices buried in the protein core or in cellular membranes have a higher and more regular distribution of hydrophobic amino acids, and are highly predictive of such structures. Helices exposed on the surface have a lower proportion of hydrophobic amino acids. Amino acid content can be predictive of an α -helical region. Regions richer in alanine (A), glutamic acid (E), leucine (L), and methionine (M) and poorer in proline (P), glycine (G), tyrosine (Y), and serine (S) tend to form an α helix. Proline destabilizes or breaks an α helix but can be present in longer helices, forming a bend.

β sheet

β sheets are formed by H bonds between an average of 5–10 consecutive amino acids in one portion of the chain with another 5–10 farther down the chain. The interacting regions may be adjacent, with a short loop in between, or far apart, with other structures in between. Every chain may run in the same direction to form a parallel sheet, every other chain may run in the reverse chemical direction to form an anti parallel sheet, or the chains may be parallel and anti parallel to form a mixed sheet.The pattern of H bonding is different in the parallel and anti parallel configurations. Each amino acid in the interior strands of the sheet forms two H bonds with neighboring amino acids, whereas each amino acid on the outside strands forms only one bond with an interior strand. Looking across the sheet at right angles to the strands, more distant strands are rotated slightly counterclockwise to form a left-handed twist. The Cα atoms alternate above and below the sheet in a pleated structure, and the R side groups of the amino acids alternate above and below the pleats. The Φ and Ψ angles of the amino acids in sheets vary considerably in one region of the Ramachandran plot. It is more difficult to predict the location of β sheets than of α helices. The situation improves somewhat when the amino acid variation in multiple sequence alignments is taken into account.

Loop

Loops are regions of a protein chain that are (1) between α helices and β sheets, (2) of various lengths and three-dimensional configurations, and (3) on the surface of the structure. Hairpin loops that represent a complete turn in the polypeptide chain joining two antiparallel β strands may be as short as two amino acids in length. Loops interact with the surrounding aqueous environment and other proteins. Because amino acids in loops are not constrained by space and environment as are amino acids in the core region, and do not have an effect on the arrangement of secondary structures in the core, more substitutions, insertions, and deletions may occur. Thus, in a sequence alignment, the presence of these features may be an indication of a loop. The positions of introns in genomic DNA sometimes correspond to the locations of loops in the encoded protein. Loops also tend to have charged and polar amino acids and are frequently a component of active sites. A detailed examination of loop structures has shown that they fall into distinct families.

Coils

A region of secondary structure that is not a α helix, a β sheet, or a recognizable turn is commonly referred to as a coil.

Protein Classification

Proteins may be classified according to both structural and sequence similarity. For structural classification, the sizes and spatial arrangements of secondary structures described in the above paragraph are compared in known three-dimensional structures.Classification based on sequence similarity was historically the first to be used. Initially, similarity based on alignments of whole sequences was performed. Later, proteins were classified on the basis of the occurrence of conserved amino acid patterns. Databases that classify proteins by one or more of these schemes are available. In considering protein classification schemes, it is important to keep several observations in mind. First, two entirely different protein sequences from different evolutionary origins may fold into a similar structure. Conversely, the sequence of an ancient gene for a given structure may have diverged considerably in different species while at the same time maintaining the same basic structural features. Recognizing any remaining sequence similarity in such cases may be a very difficult task. Second, two proteins that share a significant degree of sequence similarity either with each other or with a third sequence also share an evolutionary origin and should share some structural features also. However, gene duplication and genetic rearrangements during evolution may give rise to new gene copies, which can then evolve into proteins with new function and structure.

Terms used for Classifying Protein Structures and Sequences

The more commonly used terms for evolutionary and structural relationships among proteins are listed below. Many additional terms are used for various kinds of structural features found in proteins. Descriptions of such terms may be found at the CATH Web site the Structural Classification of Proteins (SCOP) Web site and a Glaxo-Wellcome tutorial on the Swiss bioinformatics Expasy Web site.

active site

> a localized combination of amino acid side groups within the tertiary (three-dimensional) or quaternary (protein subunit) structure that can interact with a chemically specific substrate and that provides the protein with biological activity. Proteins of very different amino acid sequences may fold into a structure that produces the same active site.

architecture

> the relative orientations of secondary structures in a three-dimensional structure without regard to whether or not they share a similar loop structure.

fold

> a type of architecture that also has a conserved loop structure.

blocks

> a conserved amino acid sequence pattern in a family of proteins. The pattern includes a series of possible matches at each position in the rep- resented sequences, but there are not any inserted or deleted positions in the pattern or in the sequences. By way of contrast, sequence profiles are a type of scoring matrix that represents a similar set of patterns that includes insertions and deletions.

class

a term used to classify protein domains according to their secondary structural content and organization. Four classes were originally recognized by Levitt and Chothia (1976), and several others have been added in the SCOP database. Three classes are given in the CATH database: mainly-α, mainly-β, and α–β, with the α–β class including both alternating α/β and α+β structures.

core

the portion of a folded protein molecule that comprises the hydrophobic interior of α-helices and β-sheets. The compact structure brings together side groups of amino acids into close enough proximity so that they can interact. When comparing protein structures, as in the SCOP database, core is the region common to most of the structures that share a common fold or that are in the same superfamily. In structure prediction, core is sometimes defined as the arrangement of secondary structures that is likely to be conserved during evolutionary change.

domain (sequence context)

a segment of a polypeptide chain that can fold into a three-dimensional structure irrespective of the presence of other segments of the chain. The separate domains of a given protein may interact extensively or may be joined only by a length of polypeptide chain. A protein with several domains may use these domains for functional interactions with different molecules.

family (sequence context)

a group of proteins of similar biochemical function that are more than 50% identical when aligned. This same cutoff is still used by the Protein Information Resource (PIR). A protein family comprises proteins with the same function in different organisms (orthologous sequences) but may also include proteins in the same organism (paralogous sequences) derived from gene duplication and rearrangements. If a multiple sequence alignment of a protein family reveals a common level of similarity throughout the lengths of the proteins, PIR refers to the family as a homeomorphic family. The aligned region is referred to as a homeomorphic domain, and this region may comprise several smaller homology domains that are shared with other families. Families may be further subdivided into subfamilies or grouped into superfamilies based on respective higher or lower levels of sequence similarity. The SCOP database reports 1296 families and the CATH database (version 1.7 beta), reports 1846 families.

When the sequences of proteins with the same function are examined in greater detail, some are found to share high sequence similarity. They are obviously members of the same family by the above criteria. However, others are found that have very little, or even insignificant, sequence similarity with other family members. In such cases, the family relationship between two distant family members A and C can often be demonstrated by finding an additional family member B that shares significant similarity with both A and C. Thus, B provides a connecting link between A and C. Another approach is to examine distant alignments for highly conserved matches.

At a level of identity of 50%, proteins are likely to have the same three-dimensional structure, and the identical atoms in the sequence alignment will also superimpose within approximately 1 Å in the structural model. Thus, if the structure of one member of a family is known, a reliable prediction may be made for a second member of the family, and the higher the identity level, the more reliable the prediction. Protein structural modeling can be performed by examining how well the amino acid substitutions fit into the core of the three-dimensional structure.

family (structural context)

as used in the FSSP database (Families of structurally similar proteins) and the DALI/FSSP Web site, two structures that have a significant level of structural similarity but not necessarily significant sequence similarity.

fold

similar to structural motif, includes a larger combination of secondary structural units in the same configuration. Thus, proteins sharing the same fold have the same combination of secondary structures that are connected by similar loops. An example is the Rossman fold comprising several alternating α helices and parallel β strands. In the SCOP, CATH, and FSSP databases, the known protein structures have been classified into hierarchical levels of structural complexity with the fold as a basic level of classification.

homologous domain (sequence context)

an extended sequence pattern, generally found by sequence alignment methods, that indicates a common evolutionary origin among the aligned sequences. A homology domain is generally longer than motifs. The domain may include all of a given protein sequence or only a portion of the sequence. Some domains are complex and made up of several smaller homology domains that became joined to form a larger one during evolution. A domain that covers an entire sequence is called the homeomorphic domain by PIR (Protein Information Resource).

module

a region of conserved amino acid patterns comprising one or more motifs and considered to be a fundamental unit of structure or function. The presence of a module has also been used to classify proteins into families.

motif (sequence context)

a conserved pattern of amino acids that is found in two or more proteins. In the Prosite catalog, a motif is an amino acid pattern that is found in a group of proteins that have a similar biochemical activity, and that often is near the active site of the protein. Examples of sequence motif databases are the Prosite catalog and the Stanford Motifs Database.

motif (structural context)

a combination of several secondary structural elements produced by the folding of adjacent sections of the polypeptide chain into a specific three-dimensional configuration. An example is the helix-loop-helix motif. Structural motifs are also referred to as supersecondary structures and folds.

position-specific scoring matrix (sequence context, also known as weight or scoring matrix)

represents a conserved region in a multiple sequence alignment with no gaps. Each matrix column represents the variation found in one column of the multiple sequence alignment.

Position-specific scoring matrix—3D (structural context) represents the amino acid variation found in an alignment of proteins that fall into the same structural class. Matrix columns represent the amino acid variation found at one amino acid position in the aligned structures.

primary structure

the linear amino acid sequence of a protein, which chemically is a polypeptide chain composed of amino acids joined by peptide bonds.

profile (sequence context)

a scoring matrix that represents a multiple sequence alignment of a protein family. The profile is usually obtained from a well-conserved region in a multiple sequence alignment. The profile is in the form of a matrix with each column representing a position in the alignment and each row one of the amino acids. Matrix values give the likelihood of each amino acid at the corresponding position in the alignment. The profile is moved along the target sequence to locate the best scoring regions by a dynamic programming algorithm. Gaps are allowed during matching and a gap penalty is included in this case as a negative score when no amino acid is matched. A sequence profile may also be represented by a hidden Markov model, referred to as a profile HMM (hidden markov model).

profile (structural context)

a scoring matrix that represents which amino acids should fit well and which should fit poorly at sequential positions in a known protein structure. Profile columns represent sequential positions in the structure, and profile rows represent the 20 amino acids. As with a sequence profile, the structural profile is moved along a target sequence to find the highest possible alignment score by a dynamic programming algorithm. Gaps may be included and receive a penalty. The resulting score provides an indication as to whether or not the target protein might adopt such a structure.

quaternary structure

the three-dimensional configuration of a protein molecule comprising several independent polypeptide chains.

secondary structure

the interactions that occur between the C, O, and NH groups on amino acids in a polypeptide chain to form α-helices, β-sheets, turns, loops, and other forms, and that facilitate the folding into a three-dimensional structure.

superfamily

a group of protein families of the same or different lengths that are related by distant yet detectable sequence similarity. Members of a given superfamily thus have a common evolutionary origin. Originally, Dayhoff defined the cutoff for superfamily status as being the chance that the sequences are not related of 10 6, on the basis of an alignment score (Dayhoff et al. 1978). Proteins with few identities in an alignment of the sequences but with a convincingly common number of structural and functional features are placed in the same superfamily. At the level of three-dimensional structure, superfamily proteins will share common structural features such as a common fold, but there may also be differences in the number and arrangement of secondary structures. The PIR resource uses the term *homeomorphic superfamilies* to refer to superfamilies that are composed of sequences that can be aligned from end to end, representing a sharing of single sequence homology domain, a region of similarity that extends throughout the alignment. This domain may also comprise smaller homology domains that are shared with other protein families and superfamilies. Although a given protein sequence may contain domains found in several superfamilies, thus indicating a complex evolutionary history, sequences will be assigned to only one homeomorphic superfamily based on the presence of similarity throughout a multiple sequence alignment. The superfamily alignment may also include regions that do not align either within or at the ends of the alignment. In contrast, sequences in the same family align well throughout the alignment.

supersecondary structure

a term with similar meaning to a structural motif. Tertiary structure is the three-dimensional or globular structure formed by the packing together or folding of secondary structures of a polypeptide chain.

Secondary Structure

Secondary structure prediction is a set of techniques in bioinformatics that aim to predict the local secondary structures of proteins based only on knowledge of their amino acid sequence. For proteins, a prediction consists of assigning regions of the amino acid sequence as likely alpha helices, beta strands (often noted as "extended" conformations), or turns. The success of a prediction is determined by comparing it to the results of the DSSP algorithm (or similar e.g. STRIDE) applied to the crystal structure of the protein. Specialized algorithms have been developed for the detection of specific well-defined patterns such as transmembrane helices and coiled coils in proteins.

The best modern methods of secondary structure prediction in proteins reach about 80% accuracy; this high accuracy allows the use of the predictions as feature improving fold recognition and ab initio protein structure prediction, classification of structural motifs, and refinement of sequence alignments. The accuracy of current protein secondary structure prediction methods is assessed in weekly benchmarks such as LiveBench and EVA.

Background

Early methods of secondary structure prediction, introduced in the 1960s and early 1970s, focused on identifying likely alpha helices and were based mainly on helix-coil transition models.

Significantly more accurate predictions that included beta sheets were introduced in the 1970s and relied on statistical assessments based on probability parameters derived from known solved structures. These methods, applied to a single sequence, are typically at most about 60-65% accurate, and often underpredict beta sheets. The evolutionary conservation of secondary structures can be exploited by simultaneously assessing many homologous sequences in a multiple sequence alignment, by calculating the net secondary structure propensity of an aligned column of amino acids. In concert with larger databases of known protein structures and modern machine learning methods such as neural nets and support vector machines, these methods can achieve up 80% overall accuracy in globular proteins. The theoretical upper limit of accuracy is around 90%, partly due to idiosyncrasies in DSSP assignment near the ends of secondary structures, where local conformations vary under native conditions but may be forced to assume a single conformation in crystals due to packing constraints. Limitations are also imposed by secondary structure prediction's inability to account for tertiary structure; for example, a sequence predicted as a likely helix may still be able to adopt a beta-strand conformation if it is located within a beta-sheet region of the protein and its side chains pack well with their neighbors. Dramatic conformational changes related to the protein's function or environment can also alter local secondary structure.

Historical Perspective

To date, over 20 different secondary structure prediction methods have been developed. One of the first algorithms was Chou-Fasman method, which relies predominantly on probability parameters determined from relative frequencies of each amino acid's appearance in each type of secondary structure. The original Chou-Fasman parameters, determined from the small sample of structures solved in the mid-1970s, produce poor results compared to modern methods, though the parameterization has been updated since it was first published. The Chou-Fasman method is roughly 50-60% accurate in predicting secondary structures.

The next notable program was the GOR method, named for the three scientists who developed it — *G*arnier, *O*sguthorpe, and *R*obson, is an information theory-based method. It uses the more powerful probabilistic technique of Bayesian inference. The GOR method takes into account not only the probability of each amino acid having a particular secondary structure, but also the conditional probability of the amino acid assuming each structure given the contributions of its neighbors (it does not assume that the neighbors have that same structure). The approach is both more sensitive and more accurate than that of Chou and Fasman because amino acid structural propensities are only strong for a small number of amino acids such as proline and glycine. Weak contributions from each of many neighbors can add up to strong effects overall. The original GOR method was roughly 65% accurate and is dramatically more successful in predicting alpha helices than beta sheets, which it frequently mispredicted as loops or disorganized regions.

Another big step forward, was using machine learning methods. First artificial neural networks methods were used. As a training sets they use solved structures to identify common sequence motifs associated with particular arrangements of secondary structures. These methods are over 70% accurate in their predictions, although beta strands are still often underpredicted due to the lack of three-dimensional structural information that would allow assessment of hydrogen bonding patterns that can promote formation of the extended conformation required for the presence of a complete beta sheet. PSIPRED and JPRED are some of the most known programs based on neural networks for protein secondary structure prediction. Next, support vector machines have proven particularly useful for predicting the locations of turns, which are difficult to identify with statistical methods.

Extensions of machine learning techniques attempt to predict more fine-grained local properties of proteins, such as backbone dihedral angles in unassigned regions. Both SVMs and neural networks have been applied to this problem. More recently, real-value torsion angles can be accurately predicted by SPINE-X and successfully employed for ab initio structure prediction.

Other Improvements

It is reported that in addition to the protein sequence, secondary structure formation depends on other factors. For example, it is reported that secondary structure tendencies depend also on local environment, solvent accessibility of residues, protein structural class, and even the organism from which the proteins are obtained. Based on such observations, some studies have shown that secondary structure prediction can be improved by addition of information about protein structural class, residue accessible surface area and also contact number information.

Tertiary Structure

The practical role of protein structure prediction is now more important than ever. Massive amounts of protein sequence data are produced by modern large-scale DNA sequencing efforts such as the Human Genome Project. Despite community-wide efforts in structural genomics, the output of experimentally determined protein structures—typically by time-consuming and relatively expensive X-ray crystallography or NMR spectroscopy—is lagging far behind the output of protein sequences.

The protein structure prediction remains an extremely difficult and unresolved undertaking. The two main problems are calculation of protein free energy and finding the global minimum of this energy. A protein structure prediction method must explore the space of possible protein structures which is astronomically large. These problems can be partially bypassed in "comparative" or homology modeling and fold recognition methods, in which the search space is pruned by the assumption that the protein in question adopts a structure that is close to the experimentally determined structure of another homologous protein. On the other hand, the *de novo* or ab initio protein structure prediction methods must explicitly resolve these problems. The progress and challenges in protein structure prediction has been reviewed in Zhang 2008.

Ab initio Protein Modelling

Energy- and Fragment-based Methods

Ab initio- or *de novo*- protein modelling methods seek to build three-dimensional protein models "from scratch", i.e., based on physical principles rather than (directly) on previously solved structures. There are many possible procedures that either attempt to mimic protein folding or apply some stochastic method to search possible solutions (i.e., global optimization of a suitable energy function). These procedures tend to require vast computational resources, and have thus only been carried out for tiny proteins. To predict protein structure *de novo* for larger proteins will require better algorithms and larger computational resources like those afforded by either powerful supercomputers (such as Blue Gene or MDGRAPE-3) or distributed computing (such as Folding@ home, the Human Proteome Folding Project and Rosetta@Home). Although these computational barriers are vast, the potential benefits of structural genomics (by predicted or experimental methods) make *ab initio* structure prediction an active research field.

As of 2009, a 50-residue protein could be simulated atom-by-atom on a supercomputer for 1 millisecond. As of 2012, comparable stable-state sampling could be done on a standard desktop with a new graphics card and more sophisticated algorithms. A much larger simulation timescales can be achieved using coarse-grained modeling.

Evolutionary Covariation to Predict 3D Contacts

As sequencing became more commonplace in the 1990s several groups used protein sequence alignments to predict correlated mutations and it was hoped that these coevolved residues could be used to predict tertiary structure (using the analogy to distance constraints from experimental procedures such as NMR). The assumption is when single residue mutations are slightly deleterious, compensatory mutations may occur to restabilize residue-residue interactions. This early work used what are known as *local* methods to calculate correlated mutations from protein sequences, but suffered from indirect false correlations which result from treating each pair of residues as independent of all other pairs.

In 2011, a different, and this time *global* statistical approach, demonstrated that predicted coevolved residues were sufficient to predict the 3D fold of a protein, providing there are enough sequences available (>1,000 homologous sequences are needed). The method, EVfold, uses no homology modeling, threading or 3D structure fragments and can be run on a standard personal computer even for proteins with hundreds of residues. The accuracy of the contacts predicted using this and related approaches has now been demonstrated on many known structures and contact maps, including the prediction of experimentally unsolved transmembrane proteins.

Comparative Protein Modeling

Comparative protein modelling uses previously solved structures as starting points, or templates. This is effective because it appears that although the number of actual proteins is vast, there is a limited set of tertiary structural motifs to which most proteins belong. It has been suggested that there are only around 2,000 distinct protein folds in nature, though there are many millions of different proteins.

These methods may also be split into two groups:

Homology modeling

> is based on the reasonable assumption that two homologous proteins will share very similar structures. Because a protein's fold is more evolutionarily conserved than its amino acid sequence, a target sequence can be modeled with reasonable accuracy on a very distantly related template, provided that the relationship between target and template can be discerned through sequence alignment. It has been suggested that the primary bottleneck in comparative modelling arises from difficulties in alignment rather than from errors in

structure prediction given a known-good alignment. Unsurprisingly, homology modelling is most accurate when the target and template have similar sequences.

Protein threading

scans the amino acid sequence of an unknown structure against a database of solved structures. In each case, a scoring function is used to assess the compatibility of the sequence to the structure, thus yielding possible three-dimensional models. This type of method is also known as 3D-1D fold recognition due to its compatibility analysis between three-dimensional structures and linear protein sequences. This method has also given rise to methods performing an inverse folding search by evaluating the compatibility of a given structure with a large database of sequences, thus predicting which sequences have the potential to produce a given fold.

Side-chain Geometry Prediction

Accurate packing of the amino acid side chains represents a separate problem in protein structure prediction. Methods that specifically address the problem of predicting side-chain geometry include dead-end elimination and the self-consistent mean field methods. The side chain conformations with low energy are usually determined on the rigid polypeptide backbone and using a set of discrete side chain conformations known as "rotamers." The methods attempt to identify the set of rotamers that minimize the model's overall energy.

These methods use rotamer libraries, which are collections of favorable conformations for each residue type in proteins. Rotamer libraries may contain information about the conformation, its frequency, and the standard deviations about mean dihedral angles, which can be used in sampling. Rotamer libraries are derived from structural bioinformatics or other statistical analysis of side-chain conformations in known experimental structures of proteins, such as by clustering the observed conformations for tetrahedral carbons near the staggered (60°, 180°, -60°) values.

Rotamer libraries can be backbone-independent, secondary-structure-dependent, or backbone-dependent. Backbone-independent rotamer libraries make no reference to backbone conformation, and are calculated from all available side chains of a certain type (for instance, the first example of a rotamer library, done by Ponder and Richards at Yale in 1987). Secondary-structure-dependent libraries present different dihedral angles and/or rotamer frequencies for α – helix, β -sheet, or coil secondary structures. Backbone-dependent rotamer libraries present conformations and/or frequencies dependent on the local backbone conformation as defined by the backbone dihedral angles ϕ and ψ, regardless of secondary structure.

The modern versions of these libraries as used in most software are presented as multidimensional distributions of probability or frequency, where the peaks correspond to the dihedral-angle conformations considered as individual rotamers in the lists. Some versions are based on very carefully curated data and are used primarily for structure validation, while others emphasize relative frequencies in much larger data sets and are the form used primarily for structure prediction, such as the Dunbrack rotamer libraries.

Side-chain packing methods are most useful for analyzing the protein's hydrophobic core, where side chains are more closely packed; they have more difficulty addressing the looser constraints and higher flexibility of surface residues, which often occupy multiple rotamer conformations rather than just one.

Prediction of Structural Classes

Statistical methods have been developed for predicting structural classes of proteins based on their amino acid composition, pseudo amino acid composition and functional domain composition.

Quaternary Structure

In the case of complexes of two or more proteins, where the structures of the proteins are known or can be predicted with high accuracy, protein–protein docking methods can be used to predict the structure of the complex. Information of the effect of mutations at specific sites on the affinity of the complex helps to understand the complex structure and to guide docking methods.

Software

A great number of software tools for protein structure prediction exist. Approaches include homology modeling, protein threading, *ab initio* methods, secondary structure prediction, and transmembrane helix and signal peptide prediction. Two most successful methods based on CASP experiment are I-TASSER and HHpred.

Evaluation of Automatic Structure Prediction Servers

CASP, which stands for Critical Assessment of Techniques for Protein Structure Prediction, is a community-wide experiment for protein structure prediction taking place every two years since 1994. CASP provides with an opportunity to assess the quality of available human, non-automated methodology (human category) and automatic servers for protein structure prediction (server category, introduced in the CASP7). The official results of automated assessment in 2012 CASP10 are available at for automated servers and for human and server predictors. In December 2014 next CASP11 assessment will be publicly available.

The CAMEO3D Continuous Automated Model EvaluatiOn Server evaluates automated protein structure prediction servers on a weekly basis using blind predictions for newly release protein structures. CAMEO publishes the results on its website ().

Enzyme

Enzymes are macromolecular biological catalysts. Enzymes accelerate, or catalyze, chemical reactions. The molecules at the beginning of the process upon which enzymes may act are called substrates and the enzyme converts these into different molecules, called products. Almost all metabolic processes in the cell need enzymes in order to occur at rates fast enough to sustain life. The set of enzymes made in a cell determines which metabolic pathways occur in that cell. The study of enzymes is called *enzymology*.

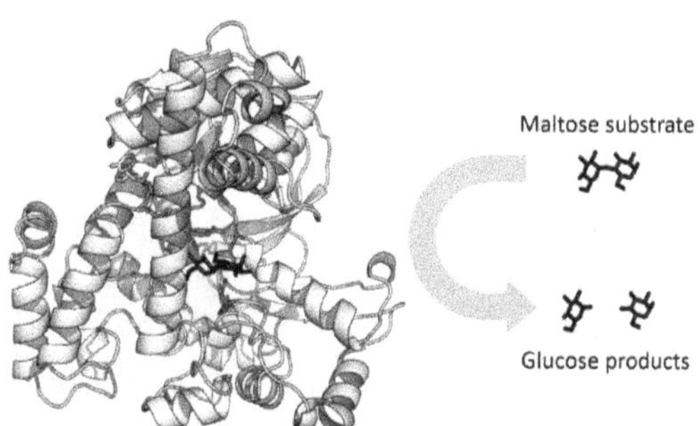

The enzyme glucosidase converts sugar maltose to two glucose sugars. Active site residues in red, maltose substrate in black, and NAD cofactor in yellow. (PDB: 1OBB)

Enzymes are known to catalyze more than 5,000 biochemical reaction types. Most enzymes are proteins, although a few are catalytic RNA molecules. Enzymes' specificity comes from their unique three-dimensional structures.

Like all catalysts, enzymes increase the rate of a reaction by lowering its activation energy. Some enzymes can make their conversion of substrate to product occur many millions of times faster. An extreme example is orotidine 5'-phosphate decarboxylase, which allows a reaction that would otherwise take millions of years to occur in milliseconds. Chemically, enzymes are like any catalyst and are not consumed in chemical reactions, nor do they alter the equilibrium of a reaction. Enzymes differ from most other catalysts by being much more specific. Enzyme activity can be affected by other molecules: inhibitors are molecules that decrease enzyme activity, and activators are molecules that increase activity. Many drugs and poisons are enzyme inhibitors. An enzyme's activity decreases markedly outside its optimal temperature and pH.

Some enzymes are used commercially, for example, in the synthesis of antibiotics. Some household products use enzymes to speed up chemical reactions: enzymes in biological washing powders break down protein, starch or fat stains on clothes, and enzymes in meat tenderizer break down proteins into smaller molecules, making the meat easier to chew.

Etymology and History

Eduard Buchner

By the late 17th and early 18th centuries, the digestion of meat by stomach secretions and the conversion of starch to sugars by plant extracts and saliva were known but the mechanisms by which these occurred had not been identified.

French chemist Anselme Payen was the first to discover an enzyme, diastase, in 1833. A few decades later, when studying the fermentation of sugar to alcohol by yeast, Louis Pasteur concluded that this fermentation was caused by a vital force contained within the yeast cells called "ferments", which were thought to function only within living organisms. He wrote that "alcoholic fermentation is an act correlated with the life and organization of the yeast cells, not with the death or putrefaction of the cells."

In 1877, German physiologist Wilhelm Kühne (1837–1900) first used the term *enzyme*, which comes from "leavened", to describe this process. The word *enzyme* was used later to refer to nonliving substances such as pepsin, and the word *ferment* was used to refer to chemical activity produced by living organisms.

Eduard Buchner submitted his first paper on the study of yeast extracts in 1897. In a series of experiments at the University of Berlin, he found that sugar was fermented by yeast extracts even when there were no living yeast cells in the mixture. He named the enzyme that brought about the fermentation of sucrose "zymase". In 1907, he received the Nobel Prize in Chemistry for "his discovery of cell-free fermentation". Following Buchner's example, enzymes are usually named according to the reaction they carry out: the suffix *-ase* is combined with the name of the substrate (e.g., lactase is the enzyme that cleaves lactose) or to the type of reaction (e.g., DNA polymerase forms DNA polymers).

The biochemical identity of enzymes was still unknown in the early 1900s. Many scientists observed that enzymatic activity was associated with proteins, but others (such as Nobel laureate Richard Willstätter) argued that proteins were merely carriers for the true enzymes and that proteins *per se* were incapable of catalysis. In 1926, James B. Sumner showed that the enzyme urease was a pure protein and crystallized it; he did likewise for the enzyme catalase in 1937. The conclusion that pure proteins can be enzymes was definitively demonstrated by John Howard Northrop and Wendell Meredith Stanley, who worked on the digestive enzymes pepsin (1930), trypsin and chymotrypsin. These three scientists were awarded the 1946 Nobel Prize in Chemistry.

The discovery that enzymes could be crystallized eventually allowed their structures to be solved by x-ray crystallography. This was first done for lysozyme, an enzyme found in tears, saliva and egg whites that digests the coating of some bacteria; the structure was solved by a group led by David Chilton Phillips and published in 1965. This high-resolution structure of lysozyme marked the beginning of the field of structural biology and the effort to understand how enzymes work at an atomic level of detail.

Naming Conventions

An enzyme's name is often derived from its substrate or the chemical reaction it catalyzes, with the word ending in *-ase*. Examples are lactase, alcohol dehydrogenase and DNA polymerase. Different enzymes that catalyze the same chemical reaction are called isozymes.

The International Union of Biochemistry and Molecular Biology have developed a nomenclature for enzymes, the EC numbers; each enzyme is described by a sequence of four numbers preceded by "EC". The first number broadly classifies the enzyme based on its mechanism.

The top-level classification is:

- EC 1, Oxidoreductases: catalyze oxidation/reduction reactions
- EC 2, Transferases: transfer a functional group (*e.g.* a methyl or phosphate group)
- EC 3, Hydrolases: catalyze the hydrolysis of various bonds
- EC 4, Lyases: cleave various bonds by means other than hydrolysis and oxidation
- EC 5, Isomerases: catalyze isomerization changes within a single molecule
- EC 6, Ligases: join two molecules with covalent bonds.

These sections are subdivided by other features such as the substrate, products, and chemical mechanism. An enzyme is fully specified by four numerical designations. For example, hexokinase (EC 2.7.1.1) is a transferase (EC 2) that adds a phosphate group (EC 2.7) to a hexose sugar, a molecule containing an alcohol group (EC 2.7.1).

Structure

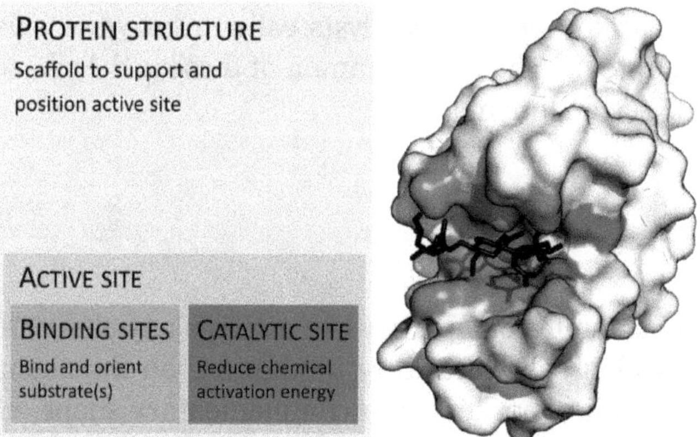

Organisation of enzyme structure and lysozyme example. Binding sites in blue, catalytic site in red and peptidoglycan substrate in black. (PDB: 9LYZ)

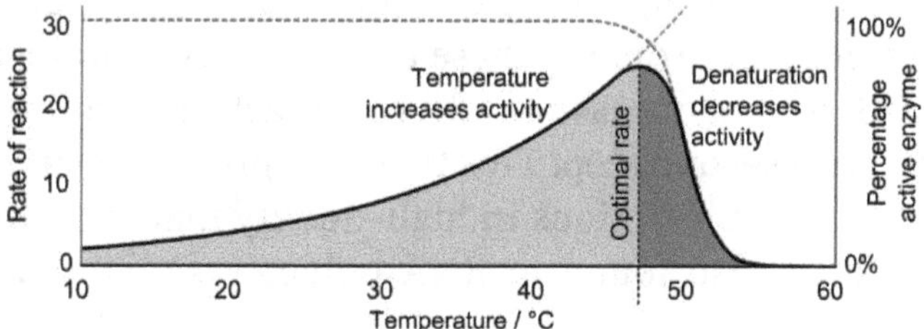

Enzyme activity initially increases with temperature (Q10 coefficient) until the enzyme's structure unfolds (denaturation), leading to an optimal rate of reaction at an intermediate temperature.

Enzymes are generally globular proteins, acting alone or in larger complexes. Like all proteins, enzymes are linear chains of amino acids that fold to produce a three-dimensional structure. The sequence of the amino acids specifies the structure which in turn determines the catalytic activity of the enzyme. Although structure determines function, a novel enzyme's activity cannot yet be predicted from its structure alone. Enzyme structures unfold (denature) when heated or exposed to chemical denaturants and this disruption to the structure typically causes a loss of activity. Enzyme denaturation is normally linked to temperatures above a species' normal level; as a result, enzymes from bacteria living in volcanic environments such as hot springs are prized by industrial users for their ability to function at high temperatures, allowing enzyme-catalysed reactions to be operated at a very high rate.

Enzymes are usually much larger than their substrates. Sizes range from just 62 amino acid residues, for the monomer of 4-oxalocrotonate tautomerase, to over 2,500 residues in the animal fatty acid synthase. Only a small portion of their structure (around 2–4 amino acids) is directly involved in catalysis: the catalytic site. This catalytic site is located next to one or more binding sites where residues orient the substrates. The catalytic site and binding site together comprise the enzyme's active site. The remaining majority of the enzyme structure serves to maintain the precise orientation and dynamics of the active site.

In some enzymes, no amino acids are directly involved in catalysis; instead, the enzyme contains sites to bind and orient catalytic cofactors. Enzyme structures may also contain allosteric sites where the binding of a small molecule causes a conformational change that increases or decreases activity.

A small number of RNA-based biological catalysts called ribozymes exist, which again can act alone or in complex with proteins. The most common of these is the ribosome which is a complex of protein and catalytic RNA components.

Mechanism

Substrate Binding

Enzymes must bind their substrates before they can catalyse any chemical reaction. Enzymes are usually very specific as to what substrates they bind and then the chemical reaction catalysed. Specificity is achieved by binding pockets with complementary shape, charge and hydrophilic/hydrophobic characteristics to the substrates. Enzymes can therefore distinguish between very similar substrate molecules to be chemoselective, regioselective and stereospecific.

Some of the enzymes showing the highest specificity and accuracy are involved in the copying and expression of the genome. Some of these enzymes have "proof-reading" mechanisms. Here, an enzyme such as DNA polymerase catalyzes a reaction in a first step and then checks that the product is correct in a second step. This two-step process results in average error rates of less than 1 error in 100 million reactions in high-fidelity mammalian polymerases. Similar proofreading mechanisms are also found in RNA polymerase, aminoacyl tRNA synthetases and ribosomes.

Conversely, some enzymes display enzyme promiscuity, having broad specificity and acting on a range of different physiologically relevant substrates. Many enzymes possess small side activities which arose fortuitously (i.e. neutrally), which may be the starting point for the evolutionary selection of a new function.

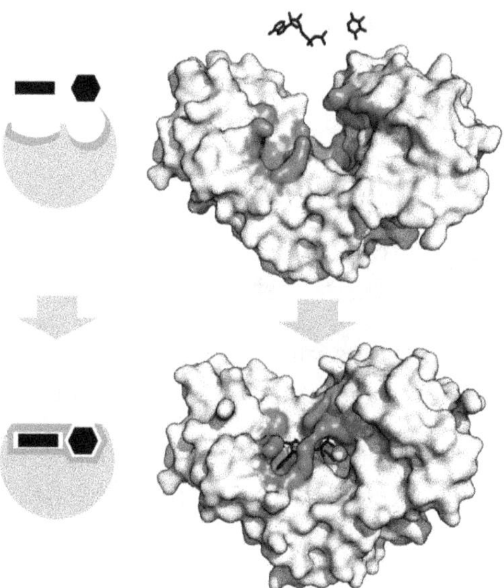

Enzyme changes shape by induced fit upon substrate binding to form enzyme-substrate complex. Hexokinase has a large induced fit motion that closes over the substrates adenosine triphosphate and xylose. Binding sites in blue, substrates in black and Mg^{2+} cofactor in yellow. (PDB: 2E2N, 2E2Q)

"Lock and key" Model

To explain the observed specificity of enzymes, in 1894 Emil Fischer proposed that both the enzyme and the substrate possess specific complementary geometric shapes that fit exactly into one another. This is often referred to as "the lock and key" model. This early model explains enzyme specificity, but fails to explain the stabilization of the transition state that enzymes achieve.

Induced Fit Model

In 1958, Daniel Koshland suggested a modification to the lock and key model: since enzymes are rather flexible structures, the active site is continuously reshaped by interactions with the substrate as the substrate interacts with the enzyme. As a result, the substrate does not simply bind to a rigid active site; the amino acid side-chains that make up the active site are molded into the precise positions that enable the enzyme to perform its catalytic function. In some cases, such as glycosidases, the substrate molecule also changes shape slightly as it enters the active site. The active site continues to change until the substrate is completely bound, at which point the final shape and charge distribution is determined. Induced fit may enhance the fidelity of molecular recognition in the presence of competition and noise via the conformational proofreading mechanism.

Catalysis

Enzymes can accelerate reactions in several ways, all of which lower the activation energy ($\Delta G^{\ddagger}$, Gibbs free energy)

1. By stabilizing the transition state:

    - Creating an environment with a charge distribution complementary to that of the transition state to lower its energy.

2. By providing an alternative reaction pathway:
    - Temporarily reacting with the substrate, forming a covalent intermediate to provide a lower energy transition state.
3. By destabilising the substrate ground state:
    - Distorting bound substrate(s) into their transition state form to reduce the energy required to reach the transition state.
    - By orienting the substrates into a productive arrangement to reduce the reaction entropy change. The contribution of this mechanism to catalysis is relatively small.

Enzymes may use several of these mechanisms simultaneously. For example, proteases such as trypsin perform covalent catalysis using a catalytic triad, stabilise charge build-up on the transition states using an oxyanion hole, complete hydrolysis using an oriented water substrate.

Dynamics

Enzymes are not rigid, static structures; instead they have complex internal dynamic motions – that is, movements of parts of the enzyme's structure such as individual amino acid residues, groups of residues forming a protein loop or unit of secondary structure, or even an entire protein domain. These motions give rise to a conformational ensemble of slightly different structures that interconvert with one another at equilibrium. Different states within this ensemble may be associated with different aspects of an enzyme's function. For example, different conformations of the enzyme dihydrofolate reductase are associated with the substrate binding, catalysis, cofactor release, and product release steps of the catalytic cycle.

Allosteric Modulation

Allosteric sites are pockets on the enzyme, distinct from the active site, that bind to molecules in the cellular environment. These molecules then cause a change in the conformation or dynamics of the enzyme that is transduced to the active site and thus affects the reaction rate of the enzyme. In this way, allosteric interactions can either inhibit or activate enzymes. Allosteric interactions with metabolites upstream or downstream in an enzyme's metabolic pathway cause feedback regulation, altering the activity of the enzyme according to the flux through the rest of the pathway.

Cofactors

Some enzymes do not need additional components to show full activity. Others require non-protein molecules called cofactors to be bound for activity. Cofactors can be either inorganic (e.g., metal ions and iron-sulfur clusters) or organic compounds (e.g., flavin and heme). Organic cofactors can be either coenzymes, which are released from the enzyme's active site during the reaction, or prosthetic groups, which are tightly bound to an enzyme. Organic prosthetic groups can be covalently bound (e.g., biotin in enzymes such as pyruvate carboxylase).

An example of an enzyme that contains a cofactor is carbonic anhydrase, which is shown in the ribbon diagram above with a zinc cofactor bound as part of its active site. These tightly bound ions or molecules are usually found in the active site and are involved in catalysis. For example, flavin and heme cofactors are often involved in redox reactions.

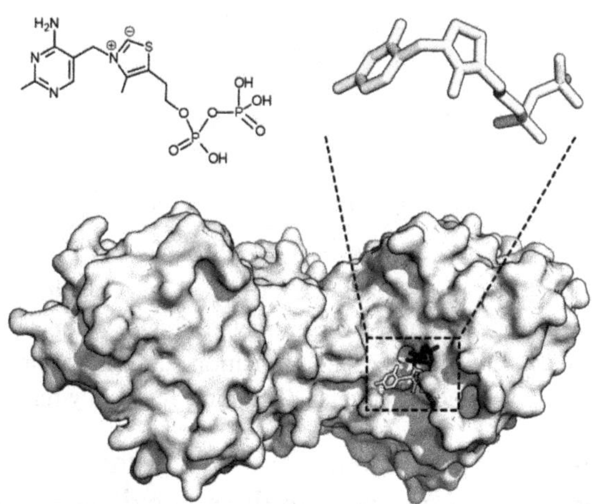

Chemical structure for thiamine pyrophosphate and protein structure of transketolase. Thiaminepyrophosphate cofactor in yellow and xylulose 5-phosphate substrate in black. (PDB: 4KXV)

Enzymes that require a cofactor but do not have one bound are called *apoenzymes* or *apoproteins*. An enzyme together with the cofactor(s) required for activity is called a *holoenzyme* (or haloenzyme). The term *holoenzyme* can also be applied to enzymes that contain multiple protein subunits, such as the DNA polymerases; here the holoenzyme is the complete complex containing all the subunits needed for activity.

Coenzymes

Coenzymes are small organic molecules that can be loosely or tightly bound to an enzyme. Coenzymes transport chemical groups from one enzyme to another. Examples include NADH, NADPH and adenosine triphosphate (ATP). Some coenzymes, such as riboflavin, thiamine and folic acid, are vitamins, or compounds that cannot be synthesized by the body and must be acquired from the diet. The chemical groups carried include the hydride ion (H^-) carried by NAD or $NADP^+$, the phosphate group carried by adenosine triphosphate, the acetyl group carried by coenzyme A, formyl, methenyl or methyl groups carried by folic acid and the methyl group carried by S-adenosylmethionine.

Since coenzymes are chemically changed as a consequence of enzyme action, it is useful to consider coenzymes to be a special class of substrates, or second substrates, which are common to many different enzymes. For example, about 1000 enzymes are known to use the coenzyme NADH.

Coenzymes are usually continuously regenerated and their concentrations maintained at a steady level inside the cell. For example, NADPH is regenerated through the pentose phosphate pathway and *S*-adenosylmethionine by methionine adenosyltransferase. This continuous regeneration means that small amounts of coenzymes can be used very intensively. For example, the human body turns over its own weight in ATP each day.

Thermodynamics

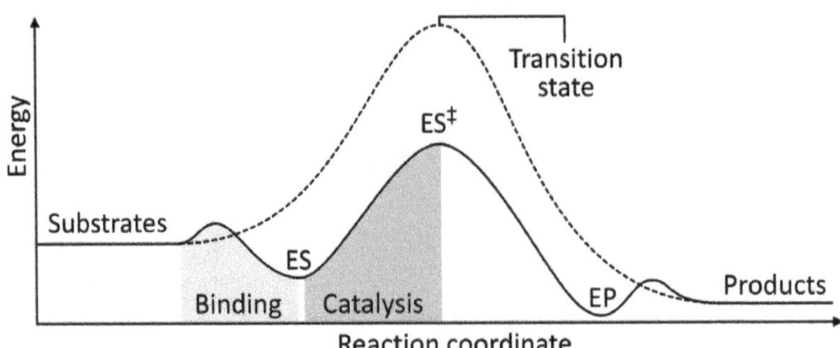

The energies of the stages of a chemical reaction. Uncatalysed (dashed line), substrates need a lot of activation energy to reach a transition state, which then decays into lower-energy products. When enzyme catalysed (solid line), the enzyme binds the substrates (ES), then stabilizes the transition state ($ES^{‡}$) to reduce the activation energy required to produce products (EP) which are finally released.

As with all catalysts, enzymes do not alter the position of the chemical equilibrium of the reaction. In the presence of an enzyme, the reaction runs in the same direction as it would without the enzyme, just more quickly. For example, carbonic anhydrase catalyzes its reaction in either direction depending on the concentration of its reactants:

$CO_2 + H_2O \xrightarrow{\text{Carbonic anhydrase}} H_2CO_3$ (in tissues; high CO_2 concentration)	(1)
$H_2CO_3 \xrightarrow{\text{Carbonic anhydrase}} CO_2 + H_2O$ (in lungs; low CO_2 concentration)	(2)

The rate of a reaction is dependent on the activation energy needed to form the transition state which then decays into products. Enzymes increase reaction rates by lowering the energy of the transition state. First, binding forms a low energy enzyme-substrate complex (ES). Secondly the enzyme stabilises the transition state such that it requires less energy to achieve compared to the uncatalyzed reaction ($ES^{‡}$). Finally the enzyme-product complex (EP) dissociates to release the products.

Enzymes can couple two or more reactions, so that a thermodynamically favorable reaction can be used to "drive" a thermodynamically unfavourable one so that the combined energy of the products is lower than the substrates. For example, the hydrolysis of ATP is often used to drive other chemical reactions.

Kinetics

Enzyme kinetics is the investigation of how enzymes bind substrates and turn them into products. The rate data used in kinetic analyses are commonly obtained from enzyme assays. In 1913 Leonor Michaelis and Maud Leonora Menten proposed a quantitative theory of enzyme kinetics, which is referred to as Michaelis–Menten kinetics. The major contribution of Michaelis and Menten was to think of enzyme reactions in two stages. In the first, the substrate binds reversibly to the enzyme, forming the enzyme-substrate complex. This is sometimes called the Michaelis-Menten complex in their honor. The enzyme then catalyzes the chemical step in the reaction and releases the product. This work was further developed by G. E. Briggs and J. B. S. Haldane, who derived kinetic equations that are still widely used today.

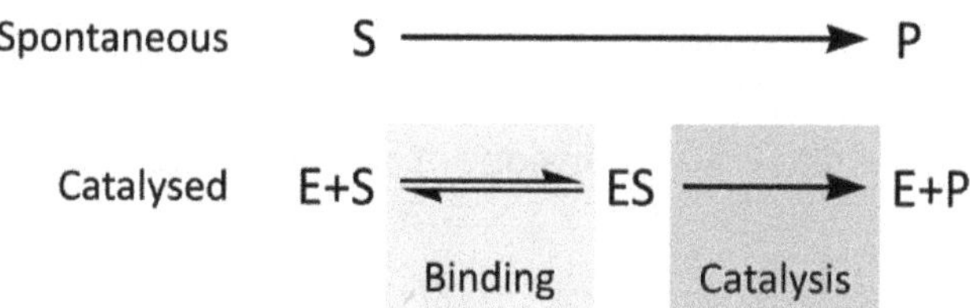

A chemical reaction mechanism with or without enzyme catalysis. The enzyme (E) binds substrate (S) to produce product (P).

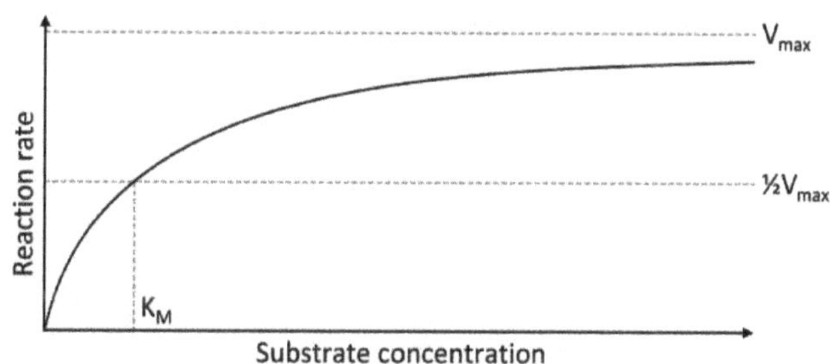

Saturation curve for an enzyme reaction showing the relation between the substrate concentration and reaction rate.

Enzyme rates depend on solution conditions and substrate concentration. To find the maximum speed of an enzymatic reaction, the substrate concentration is increased until a constant rate of product formation is seen. This is shown in the saturation curve on the right. Saturation happens because, as substrate concentration increases, more and more of the free enzyme is converted into the substrate-bound ES complex. At the maximum reaction rate (V_{max}) of the enzyme, all the enzyme active sites are bound to substrate, and the amount of ES complex is the same as the total amount of enzyme.

V_{max} is only one of several important kinetic parameters. The amount of substrate needed to achieve a given rate of reaction is also important. This is given by the Michaelis-Menten constant (K_m), which is the substrate concentration required for an enzyme to reach one-half its maximum reaction rate; generally, each enzyme has a characteristic K_m for a given substrate. Another useful constant is k_{cat}, also called the *turnover number*, which is the number of substrate molecules handled by one active site per second.

The efficiency of an enzyme can be expressed in terms of k_{cat}/K_m. This is also called the specificity constant and incorporates the rate constants for all steps in the reaction up to and including the first irreversible step. Because the specificity constant reflects both affinity and catalytic ability, it is useful for comparing different enzymes against each other, or the same enzyme with different substrates. The theoretical maximum for the specificity constant is called the diffusion limit and is about 10^8 to 10^9 (M^{-1} s^{-1}). At this point every collision of the enzyme with its substrate will result in catalysis, and the rate of product formation is not limited by the reaction rate but by the diffusion rate. Enzymes with this property are called *catalytically perfect* or *kinetically perfect*. Example of such enzymes are triose-phosphate isomerase, carbonic anhydrase, acetylcholinesterase, catalase, fumarase, β-lactamase, and superoxide dismutase. The turnover of such enzymes can reach several million reactions per second.

Michaelis–Menten kinetics relies on the law of mass action, which is derived from the assumptions of free diffusion and thermodynamically driven random collision. Many biochemical or cellular processes deviate significantly from these conditions, because of macromolecular crowding and constrained molecular movement. More recent, complex extensions of the model attempt to correct for these effects.

Inhibition

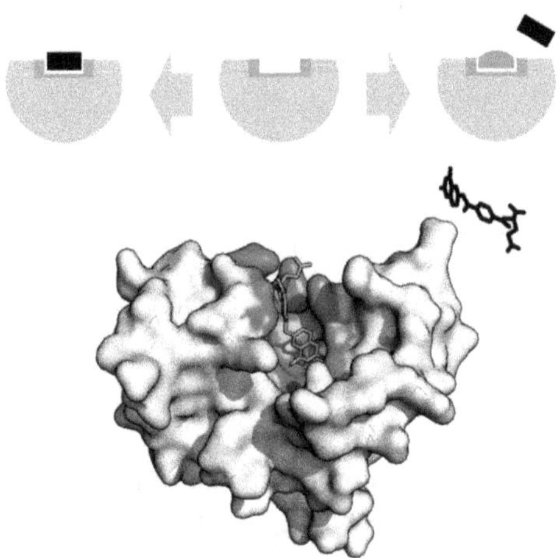

An enzyme binding site that would normally bind substrate can alternatively bind a competitive inhibitor, preventing substrate access. Dihydrofolate reductase is inhibited by methotrexate which prevents binding of its substrate, folic acid. Binding site in blue, inhibitor in green, and substrate in black. (PDB: 4QI9)

The coenzyme folic acid (left) and the anti-cancer drug methotrexate (right) are very similar in structure (differences show in green). As a result, methotrexate is a competitive inhibitor of many enzymes that use folates.

Enzyme reaction rates can be decreased by various types of enzyme inhibitors.

Types of Inhibition

Competitive

A competitive inhibitor and substrate cannot bind to the enzyme at the same time. Often competitive inhibitors strongly resemble the real substrate of the enzyme. For example, the drug methotrexate is a competitive inhibitor of the enzyme dihydrofolate reductase, which catalyzes the reduction of dihydrofolate to tetrahydrofolate. The similarity between the structures of dihydrofolate and this drug are shown in the accompanying figure. This type of inhibition can be overcome with high substrate concentration. In some cases, the inhibitor can bind to a site other than the binding-site of the usual substrate and exert an allosteric effect to change the shape of the usual binding-site.

Non-competitive

A non-competitive inhibitor binds to a site other than where the substrate binds. The substrate still binds with its usual affinity and hence K_m remains the same. However the inhibitor reduces the catalytic efficiency of the enzyme so that V_{max} is reduced. In contrast to competitive inhibition, non-competitive inhibition cannot be overcome with high substrate concentration.

Uncompetitive

An uncompetitive inhibitor cannot bind to the free enzyme, only to the enzyme-substrate complex; hence, these types of inhibitors are most effective at high substrate concentration. In the presence of the inhibitor, the enzyme-substrate complex is inactive. This type of inhibition is rare.

Mixed

A mixed inhibitor binds to an allosteric site and the binding of the substrate and the inhibitor affect each other. The enzyme's function is reduced but not eliminated when bound to the inhibitor. This type of inhibitor does not follow the Michaelis-Menten equation.

Irreversible

An irreversible inhibitor permanently inactivates the enzyme, usually by forming a covalent bond to the protein. Penicillin and aspirin are common drugs that act in this manner.

Functions of Inhibitors

In many organisms, inhibitors may act as part of a feedback mechanism. If an enzyme produces too much of one substance in the organism, that substance may act as an inhibitor for the enzyme at the beginning of the pathway that produces it, causing production of the substance to slow down or stop when there is sufficient amount. This is a form of negative feedback. Major metabolic pathways such as the citric acid cycle make use of this mechanism.

Since inhibitors modulate the function of enzymes they are often used as drugs. Many such drugs are reversible competitive inhibitors that resemble the enzyme's native substrate, similar to methotrexate above; other well-known examples include statins used to treat high cholesterol, and protease inhibitors used to treat retroviral infections such as HIV. A common example of an irreversible inhibitor that is used as a drug is aspirin, which inhibits the COX-1 and COX-2 enzymes that produce the inflammation messenger prostaglandin. Other enzyme inhibitors are poisons. For example, the poison cyanide is an irreversible enzyme inhibitor that combines with the copper and iron in the active site of the enzyme cytochrome c oxidase and blocks cellular respiration.

Biological Function

Enzymes serve a wide variety of functions inside living organisms. They are indispensable for signal transduction and cell regulation, often via kinases and phosphatases. They also generate movement, with myosin hydrolyzing ATP to generate muscle contraction, and also transport cargo around the cell as part of the cytoskeleton. Other ATPases in the cell membrane are ion pumps involved in active transport. Enzymes are also involved in more exotic functions, such as luciferase generating light in fireflies. Viruses can also contain enzymes for infecting cells, such as the HIV integrase and reverse transcriptase, or for viral release from cells, like the influenza virus neuraminidase.

An important function of enzymes is in the digestive systems of animals. Enzymes such as amylases and proteases break down large molecules (starch or proteins, respectively) into smaller ones, so they can be absorbed by the intestines. Starch molecules, for example, are too large to be absorbed from the intestine, but enzymes hydrolyze the starch chains into smaller molecules such as maltose and eventually glucose, which can then be absorbed. Different enzymes digest different food substances. In ruminants, which have herbivorous diets, microorganisms in the gut produce another enzyme, cellulase, to break down the cellulose cell walls of plant fiber.

Metabolism

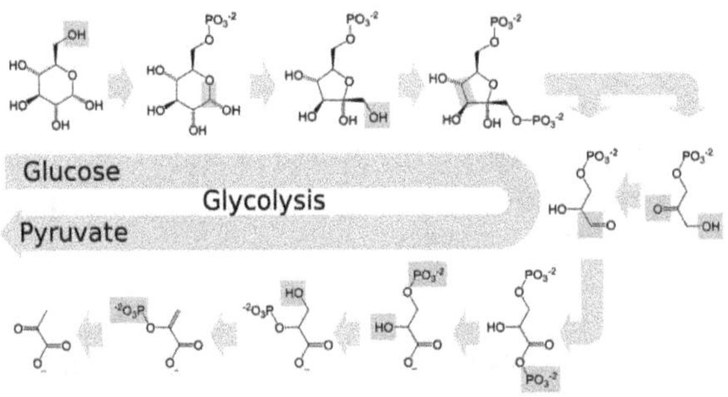

The metabolic pathway of glycolysis releases energy by converting glucose to pyruvate by via a series ofintermediate metabolites. Each chemical modification (red box) is performed by a different enzyme.

Several enzymes can work together in a specific order, creating metabolic pathways. In a metabolic pathway, one enzyme takes the product of another enzyme as a substrate. After the catalytic reaction, the product is then passed on to another enzyme. Sometimes more than one enzyme can catalyze the same reaction in parallel; this can allow more complex regulation: with, for example, a low constant activity provided by one enzyme but an inducible high activity from a second enzyme.

Enzymes determine what steps occur in these pathways. Without enzymes, metabolism would neither progress through the same steps and could not be regulated to serve the needs of the cell. Most central metabolic pathways are regulated at a few key steps, typically through enzymes whose activity involves the hydrolysis of ATP. Because this reaction releases so much energy, other reactions that are thermodynamically unfavorable can be coupled to ATP hydrolysis, driving the overall series of linked metabolic reactions.

Control of Activity

There are five main ways that enzyme activity is controlled in the cell.

Regulation

Enzymes can be either activated or inhibited by other molecules. For example, the end product(s) of a metabolic pathway are often inhibitors for one of the first enzymes of the pathway (usually the first irreversible step, called committed step), thus regulating the amount of end product made by the pathways. Such a regulatory mechanism is called a negative feedback mechanism, because the amount of the end product produced is regulated by its own concentration. Negative feedback mechanism can effectively adjust the rate of synthesis of intermediate metabolites according to the demands of the cells. This helps with effective allocations of materials and energy economy, and it prevents the excess manufacture of end products. Like other homeostatic devices, the control of enzymatic action helps to maintain a stable internal environment in living organisms.

Post-translational modification

Examples of post-translational modification include phosphorylation, myristoylation and glycosylation. For example, in the response to insulin, the phosphorylation of multiple enzymes, including glycogen synthase, helps control the synthesis or degradation of glycogen and allows the cell to respond to changes in blood sugar. Another example of post-translational modification is the cleavage of the polypeptide chain. Chymotrypsin, a digestive protease, is produced in inactive form as chymotrypsinogen in the pancreas and transported in this form to the stomach where it is activated. This stops the enzyme from digesting the pancreas or other tissues before it enters the gut. This type of inactive precursor to an enzyme is known as a zymogen or proenzyme.

Quantity

Enzyme production (transcription and translation of enzyme genes) can be enhanced or diminished by a cell in response to changes in the cell's environment. This form of gene regulation is called enzyme induction. For example, bacteria may become resistant to antibiotics such as penicillin because enzymes called beta-lactamases are induced that hydrolyse the crucial beta-lactam ring within the penicillin molecule. Another example comes from enzymes in the liver called cytochrome P450 oxidases, which are important in drug metabolism. Induction or inhibition of these enzymes can cause drug interactions. Enzyme levels can also be regulated by changing the rate of enzyme degradation.

Subcellular distribution

Enzymes can be compartmentalized, with different metabolic pathways occurring in different cellular compartments. For example, fatty acids are synthesized by one set of enzymes in the cytosol, endoplasmic reticulum and Golgi and used by a different set of enzymes as a source of energy in the mitochondrion, through β-oxidation. In addition, trafficking of the enzyme to different compartments may change the degree of protonation (cytoplasm neutral and lysosome acidic) or oxidative state [e.g., oxidized (periplasm) or reduced (cytoplasm)] which in turn affects enzyme activity.

Organ specialization

In multicellular eukaryotes, cells in different organs and tissues have different patterns of gene expression and therefore have different sets of enzymes (known as isozymes) available for metabolic reactions. This provides a mechanism for regulating the overall metabolism of the organism. For example, hexokinase, the first enzyme in the glycolysis pathway, has a specialized form called glucokinase expressed in the liver and pancreas that has a lower affinity for glucose yet is more sensitive to glucose concentration. This enzyme is involved in sensing blood sugar and regulating insulin production.

Involvement in Disease

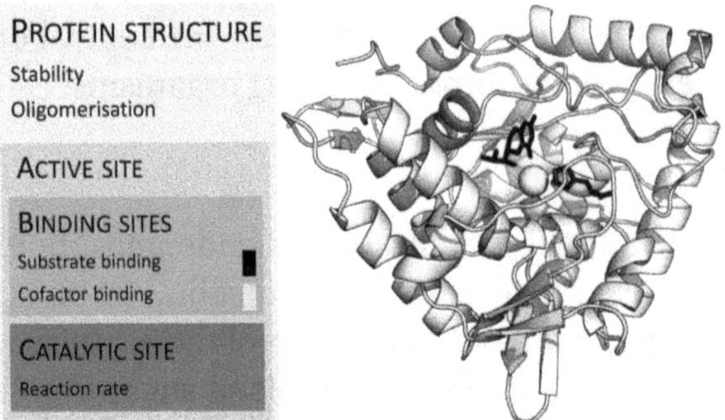

In phenylalanine hydroxylase over 300 different mutations throughout the structure cause phenylketonuria. Phenylalanine substrate and tetrahydrobiopterin coenzyme in black, and Fe^{2+} cofactor in yellow. (PDB: 1KW0)

Since the tight control of enzyme activity is essential for homeostasis, any malfunction (mutation, overproduction, underproduction or deletion) of a single critical enzyme can lead to a genetic disease. The malfunction of just one type of enzyme out of the thousands of types present in the human body can be fatal. An example of a fatal genetic disease due to enzyme insufficiency is Tay-Sachs disease, in which patients lack the enzyme hexosaminidase.

One example of enzyme deficiency is the most common type of phenylketonuria. Many different single amino acid mutations in the enzyme phenylalanine hydroxylase, which catalyzes the first step in the degradation of phenylalanine, result in build-up of phenylalanine and related products. Some mutations are in the active site, directly disrupting binding and catalysis, but many are far from the active site and reduce activity by destabilising the protein structure, or affecting correct oligomerisation. This can lead to intellectual disability if the disease is untreated. Another example is pseudocholinesterase deficiency, in which the body's ability to break down choline ester drugs is impaired. Oral administration of enzymes can be used to treat some functional enzyme deficiencies, such as pancreatic insufficiency and lactose intolerance.

Another way enzyme malfunctions can cause disease comes from germline mutations in genes coding for DNA repair enzymes. Defects in these enzymes cause cancer because cells are less able to repair mutations in their genomes. This causes a slow accumulation of mutations and results in the development of cancers. An example of such a hereditary cancer syndrome is xeroderma pigmentosum, which causes the development of skin cancers in response to even minimal exposure to ultraviolet light.

Industrial Applications

Enzymes are used in the chemical industry and other industrial applications when extremely specific catalysts are required. Enzymes in general are limited in the number of reactions they have evolved to catalyze and also by their lack of stability in organic solvents and at high temperatures. As a consequence, protein engineering is an active area of research and involves attempts to create new enzymes with novel properties, either through rational design or *in vitro* evolution. These efforts have begun to be successful, and a few enzymes have now been designed "from scratch" to catalyze reactions that do not occur in nature.

Application	Enzymes used	Uses
Biofuel industry	Cellulases	Break down cellulose into sugars that can be fermented to produce cellulosic ethanol.
	Ligninases	Pretreatment of biomass for biofuel production.
Biological detergent	Proteases, amylases, lipases	Remove protein, starch, and fat or oil stains from laundry and dishware.
	Mannanases	Remove food stains from the common food additive guar gum.
Brewing industry	Amylase, glucanases, proteases	Split polysaccharides and proteins in the malt.
	Betaglucanases	Improve the wort and beer filtration characteristics.
	Amyloglucosidase and pullulanases	Make low-calorie beer and adjust fermentability.
	Acetolactate decarboxylase (ALDC)	Increase fermentation efficiency by reducing diacetyl formation.
Culinary uses	Papain	Tenderize meat for cooking.
Dairy industry	Rennin	Hydrolyze protein in the manufacture of cheese.
	Lipases	Produce Camembert cheese and blue cheeses such as Roquefort.
Food processing	Amylases	Produce sugars from starch, such as in making high-fructose corn syrup.
	Proteases	Lower the protein level of flour, as in biscuit-making.
	Trypsin	Manufacture hypoallergenic baby foods.
	Cellulases, pectinases	Clarify fruit juices.
Molecular biology	Nucleases, DNA ligase and polymerases	Use restriction digestion and the polymerase chain reaction to create recombinant DNA.
Paper industry	Xylanases, hemicellulases and lignin peroxidases	Remove lignin from kraft pulp.
Personal care	Proteases	Remove proteins on contact lenses to prevent infections.
Starch industry	Amylases	Convert starch into glucose and various syrups.

Protein Purification

Protein purification is a series of processes intended to isolate one or a few proteins from a complex mixture, usually cells, tissues or whole organisms. Protein purification is vital for the characterization of the function, structure and interactions of the protein of interest. The purification process may separate the protein and non-protein parts of the mixture, and finally separate the desired protein from all other proteins. Separation of one protein from all others is typically the most laborious aspect of protein purification. Separation steps usually exploit differences in protein size, physico-chemical properties, binding affinity and biological activity. The pure result may be termed protein isolate.

Purpose

Protein purification is either *preparative* or *analytical*. Preparative purifications aim to produce a relatively large quantity of purified proteins for subsequent use. Examples include the preparation of commercial products such as enzymes (e.g. lactase), nutritional proteins (e.g. soy protein isolate), and certain biopharmaceuticals (e.g. insulin). Analytical purification produces a relatively small amount of a protein for a variety of research or analytical purposes, including identification, quantification, and studies of the protein's structure, post-translational modifications and function. Pepsin and urease were the first proteins purified to the point that they could be crystallized.

Recombinant bacteria can be grown in a flask containing growth media.

Preliminary Steps

Extraction

If the protein of interest is not secreted by the organism into the surrounding solution, the first step of each purification process is the disruption of the cells containing the protein. Depending on how fragile the protein is and how stable the cells are, one could, for instance, use one of the following methods: i) repeated freezing and thawing, ii) sonication, iii) homogenization by high pressure (French press), iv) homogenization by grinding (bead mill), and v) permeabilization by detergents (e.g. Triton X-100) and/or enzymes (e.g. lysozyme). Finally, the cell debris can be removed by centrifugation so that the proteins and other soluble compounds remain in the supernatant.

Also proteases are released during cell lysis, which will start digesting the proteins in the solution. If the protein of interest is sensitive to proteolysis, it is recommended to proceed quickly, and to keep the extract cooled, to slow down the digestion. Alternatively, one or more protease inhibitors can be added to the lysis buffer immediately before cell disruption. Sometimes it is also necessary to add DNAse in order to reduce the viscosity of the cell lysate caused by a high DNA content.

Precipitation and Differential Solubilization

In bulk protein purification, a common first step to isolate proteins is precipitation with ammonium sulphate $(NH_4)_2SO_4$. This is performed by adding increasing amounts of ammonium sulphate and collecting the different fractions of precipitate protein. Ammonium sulphate can be removed by dialysis.The hydrophobic groups on the proteins get exposed to the atmosphere, attract other protein hydrophobic groups and get aggregated. Protein precipitated will be large enough to be visible. One advantage of this method is that it can be performed inexpensively with very large volumes.

The first proteins to be purified are water-soluble proteins. Purification of integral membrane proteins requires disruption of the cell membrane in order to isolate any one particular protein from others that are in the same membrane compartment. Sometimes a particular membrane fraction can be isolated first, such as isolating mitochondria from cells before purifying a protein located in a mitochondrial membrane. A detergent such as sodium dodecyl sulfate (SDS) can be used to dissolve cell membranes and keep membrane proteins in solution during purification; however, because SDS causes denaturation, milder detergents such as Triton X-100 or CHAPS can be used to retain the protein's native conformation during complete purification.

Ultracentrifugation

Centrifugation is a process that uses centrifugal force to separate mixtures of particles of varying masses or densities suspended in a liquid. When a vessel (typically a tube or bottle) containing a mixture of proteins or other particulate matter, such as bacterial cells, is rotated at high speeds, the inertia of each particle yields an outward force proportional to its mass. The tendency of a given particle to move through the liquid because of this force is offset by the resistance the liquid exerts on the particle. The net effect of "spinning" the sample in a centrifuge is that massive, small, and dense particles move outward faster than less massive particles or particles with more "drag" in the liquid. When suspensions of particles are "spun" in a centrifuge, a "pellet" may form at the bottom of the vessel that is enriched for the most massive particles with low drag in the liquid.

Non-compacted particles remain mostly in the liquid called "supernatant" and can be removed from the vessel thereby separating the supernatant from the pellet. The rate of centrifugation is determined by the angular acceleration applied to the sample, typically measured in comparison to the g. If samples are centrifuged long enough, the particles in the vessel will reach equilibrium wherein the particles accumulate specifically at a point in the vessel where their buoyant density is balanced with centrifugal force. Such an "equilibrium" centrifugation can allow extensive purification of a given particle.

Sucrose gradient centrifugation — a linear concentration gradient of sugar (typically sucrose, glycerol, or a silica based density gradient media, like Percoll) is generated in a tube such that the highest concentration is on the bottom and lowest on top. Percoll is a trademark owned by GE Healthcare companies. A protein sample is then layered on top of the gradient and spun at high speeds in an ultracentrifuge. This causes heavy macromolecules to migrate towards the bottom of the tube faster than lighter material. During centrifugation in the absence of sucrose, as particles move farther and farther from the center of rotation, they experience more and more centrifugal force (the further they move, the faster they move). The problem with this is that the useful separation range of within the vessel is restricted to a small observable window. Spinning a sample twice as long doesn't mean the particle of interest will go twice as far, in fact, it will go significantly further. However, when the proteins are moving through a sucrose gradient, they encounter liquid of increasing density and viscosity. A properly designed sucrose gradient will counteract the increasing centrifugal force so the particles move in close proportion to the time they have been in the centrifugal field. Samples separated by these gradients are referred to as "rate zonal" centrifugations. After separating the protein/particles, the gradient is then fractionated and collected.

Purification Strategies

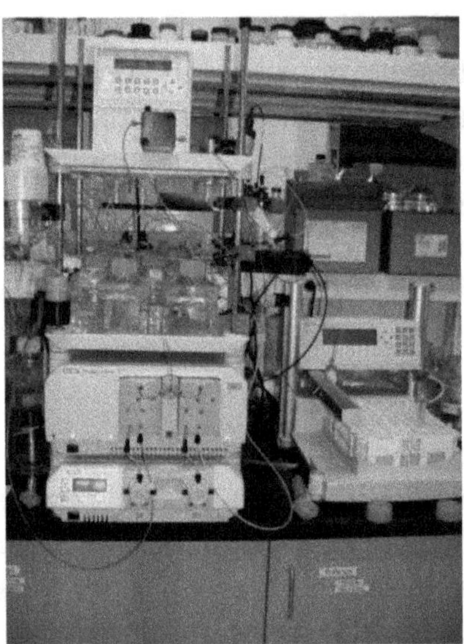

Chromatographic equipment. Here set up for a size exclusion chromatography. The buffer is pumped through the column (right) by a computer controlled device.

Choice of a starting material is key to the design of a purification process. In a plant or animal, a particular protein usually isn't distributed homogeneously throughout the body; different organs or tissues have higher or lower concentrations of the protein. Use of only the tissues or organs with the highest concentration decreases the volumes needed to produce a given amount of purified protein. If the protein is present in low abundance, or if it has a high value, scientists may use recombinant DNA technology to develop cells that will produce large quantities of the desired protein (this is known as an expression system). Recombinant expression allows the protein to be tagged, e.g. by a His-tag, to facilitate purification, which means that the purification can be done in fewer steps. In addition, recombinant expression usually starts with a higher fraction of the desired protein than is present in a natural source.

An analytical purification generally utilizes three properties to separate proteins. First, proteins may be purified according to their isoelectric points by running them through a pH graded gel or an ion exchange column. Second, proteins can be separated according to their size or molecular weight via size exclusion chromatography or by SDS-PAGE (sodium dodecyl sulfate-polyacrylamide gel electrophoresis) analysis. Proteins are often purified by using 2D-PAGE and are then analysed by peptide mass fingerprinting to establish the protein identity. This is very useful for scientific purposes and the detection limits for protein are nowadays very low and nanogram amounts of protein are sufficient for their analysis. Thirdly, proteins may be separated by polarity/hydrophobicity via high performance liquid chromatography or reversed-phase chromatography.

Usually a protein purification protocol contains one or more chromatographic steps. The basic procedure in chromatography is to flow the solution containing the protein through a column packed with various materials. Different proteins interact differently with the column material, and can thus be separated by the time required to pass the column, or the conditions required to elute the protein from the column. Usually proteins are detected as they are coming off the column by their absorbance at 280 nm. Many different chromatographic methods exist:

Size Exclusion Chromatography

Chromatography can be used to separate protein in solution or denaturing conditions by using porous gels. This technique is known as size exclusion chromatography. The principle is that smaller molecules have to traverse a larger volume in a porous matrix. Consequentially, proteins of a certain range in size will require a variable volume of eluent (solvent) before being collected at the other end of the column of gel.

In the context of protein purification, the eluent is usually pooled in different test tubes. All test tubes containing no measurable trace of the protein to purify are discarded. The remaining solution is thus made of the protein to purify and any other similarly-sized proteins.

Separation Based on Charge or Hydrophobicity

Hydrophobic Interaction Chromatography

HIC media is amphiphilic, with both hydrophobic and hydrophilic regions, allowing for separation of proteins based on their surface hydrophobicity. In pure water, the interactions between the resin and the hydrophobic regions of protein would be very weak, but this interaction is enhanced by applying a protein sample to HIC resin in high ionic strength buffer. The ionic strength of the buffer is then reduced to elute proteins in order of decreasing hydrophobicity.

Ion Exchange Chromatography

Ion exchange chromatography separates compounds according to the nature and degree of their ionic charge. The column to be used is selected according to its type and strength of charge. Anion exchange resins have a negative charge and are used to retain and separate positively charged compounds, while cation exchange resins have a positive charge and are used to separate negatively charged molecules.

Before the separation begins a buffer is pumped through the column to equilibrate the opposing charged ions. Upon injection of the sample, solute molecules will exchange with the buffer ions as each competes for the binding sites on the resin. The length of retention for each solute depends upon the strength of its charge. The most weakly charged compounds will elute first, followed by those with successively stronger charges. Because of the nature of the separating mechanism, pH, buffer type, buffer concentration, and temperature all play important roles in controlling the separation.

Ion exchange chromatography is a very powerful tool for use in protein purification and is frequently used in both analytical and preparative separations.

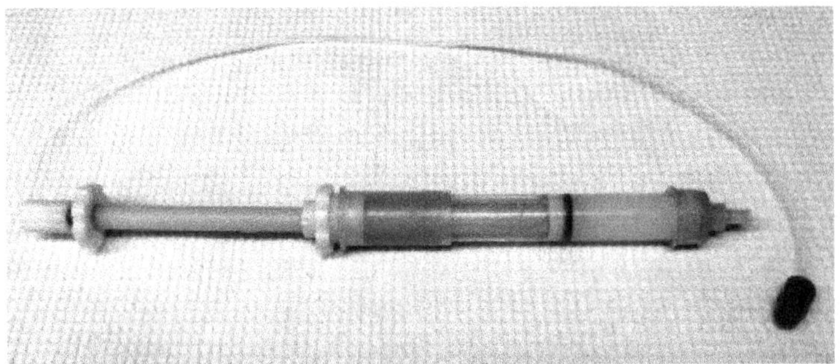

Nickel-affinity column. The resin is blue since it has bound nickel.

Free-flow-electrophoresis

Free-flow electrophoresis (FFE) is a carrier-free electrophoresis technique that allows preparative protein separation in a laminar buffer stream by using an orthogonal electric field. By making use of a pH-gradient, that can for example be induced by ampholytes, this technique allows to separate protein isoforms up to a resolution of < 0.02 delta-pI.

Affinity Chromatography

Affinity Chromatography is a separation technique based upon molecular conformation, which frequently utilizes application specific resins. These resins have ligands attached to their surfaces which are specific for the compounds to be separated. Most frequently, these ligands function in a fashion similar to that of antibody-antigen interactions. This "lock and key" fit between the ligand and its target compound makes it highly specific, frequently generating a single peak, while all else in the sample is unretained.

Many membrane proteins are glycoproteins and can be purified by lectin affinity chromatography. Detergent-solubilized proteins can be allowed to bind to a chromatography resin that has been modified to have a covalently attached lectin. Proteins that do not bind to the lectin are washed away and then specifically bound glycoproteins can be eluted by adding a high concentration of a sugar that competes with the bound glycoproteins at the lectin binding site. Some lectins have high affinity binding to oligosaccharides of glycoproteins that is hard to compete with sugars, and bound glycoproteins need to be released by denaturing the lectin.

Metal Binding

A common technique involves engineering a sequence of 6 to 8 histidines into the N- or C-terminal of the protein. The polyhistidine binds strongly to divalent metal ions such as nickel and cobalt.

The protein can be passed through a column containing immobilized nickel ions, which binds the polyhistidine tag. All untagged proteins pass through the column. The protein can be eluted with imidazole, which competes with the polyhistidine tag for binding to the column, or by a decrease in pH (typically to 4.5), which decreases the affinity of the tag for the resin. While this procedure is generally used for the purification of recombinant proteins with an engineered affinity tag (such as a 6xHis tag or Clontech's HAT tag), it can also be used for natural proteins with an inherent affinity for divalent cations.

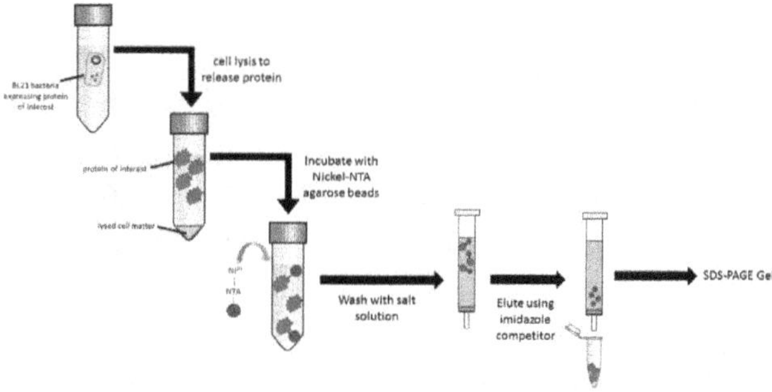

Schematic showing the steps involved in a metal binding strategy for protein purification. The use of nickel immobilized with Nitrilotriacetic acid (NTA) is shown here.

Immunoaffinity Chromatography

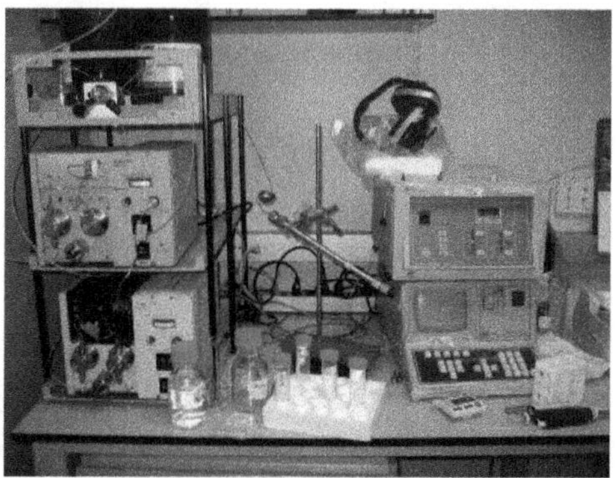

A HPLC. From left to right: A pumping device generating a gradient of two different solvents, a steel enforced column and an apparatus for measuring the absorbance.

Immunoaffinity chromatography uses the specific binding of an antibody-antigen to selectively purify the target protein. The procedure involves immobilizing a protein to a solid substrate (e.g. a porous bead or a membrane), which then selectively binds the target, while everything else flows through. The target protein can be eluted by changing the pH or the salinity. The immobilized ligand can be an antibody (such as Immunoglobulin G) or it can be a protein (such as Protein A). Because this method does not involve engineering in a tag, it can be used for proteins from natural sources.

Purification of a Tagged Protein

Another way to tag proteins is to engineer an antigen peptide tag onto the protein, and then purify the protein on a column or by incubating with a loose resin that is coated with an immobilized antibody. This particular procedure is known as immunoprecipitation. Immunoprecipitation is quite capable of generating an extremely specific interaction which usually results in binding only the desired protein. The purified tagged proteins can then easily be separated from the other proteins in solution and later eluted back into clean solution.

When the tags are not needed anymore, they can be cleaved off by a protease. This often involves engineering a protease cleavage site between the tag and the protein.

HPLC

High performance liquid chromatography or high pressure liquid chromatography is a form of chromatography applying high pressure to drive the solutes through the column faster. This means that the diffusion is limited and the resolution is improved. The most common form is "reversed phase" HPLC, where the column material is hydrophobic. The proteins are eluted by a gradient of increasing amounts of an organic solvent, such as acetonitrile. The proteins elute according to their hydrophobicity. After purification by HPLC the protein is in a solution that only contains volatile compounds, and can easily be lyophilized. HPLC purification frequently results in denaturation of the purified proteins and is thus not applicable to proteins that do not spontaneously refold.

Concentration of the Purified Protein

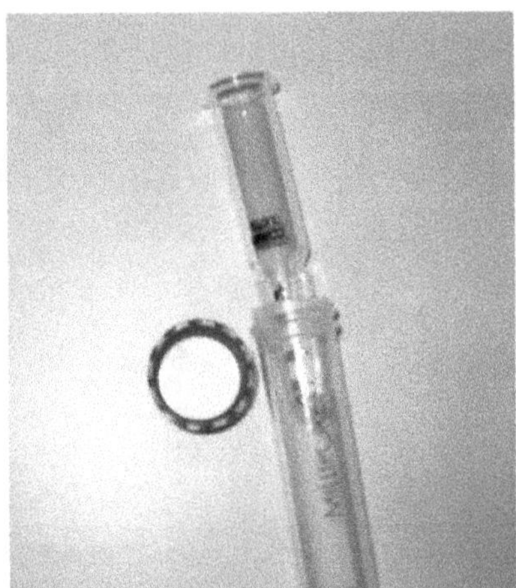

A selectively permeable membrane can be mounted in a centrifuge tube. The buffer is forced through the membrane by centrifugation, leaving the protein in the upper chamber.

At the end of a protein purification, the protein often has to be concentrated. Different methods exist.

Lyophilization

If the solution doesn't contain any other soluble component than the protein in question the protein can be lyophilized (dried). This is commonly done after an HPLC run. This simply removes all volatile components, leaving the proteins behind.

Ultrafiltration

Ultrafiltration concentrates a protein solution using selective permeable membranes. The function of the membrane is to let the water and small molecules pass through while retaining the protein. The solution is forced against the membrane by mechanical pump, gas pressure, or centrifugation.

Evaluating Purification Yield

The most general method to monitor the purification process is by running a SDS-PAGE of the different steps. This method only gives a rough measure of the amounts of different proteins in the mixture, and it is not able to distinguish between proteins with similar apparent molecular weight.

If the protein has a distinguishing spectroscopic feature or an enzymatic activity, this property can be used to detect and quantify the specific protein, and thus to select the fractions of the separation, that contains the protein. If antibodies against the protein are available then western blotting and ELISA can specifically detect and quantify the amount of desired protein. Some proteins function as receptors and can be detected during purification steps by a ligand binding assay, often using a radioactive ligand.

In order to evaluate the process of multistep purification, the amount of the specific protein has to be compared to the amount of total protein. The latter can be determined by the Bradford total protein assay or by absorbance of light at 280 nm, however some reagents used during the purification process may interfere with the quantification. For example, imidazole (commonly used for purification of polyhistidine-tagged recombinant proteins) is an amino acid analogue and at low concentrations will interfere with the bicinchoninic acid (BCA) assay for total protein quantification. Impurities in low-grade imidazole will also absorb at 280 nm, resulting in an inaccurate reading of protein concentration from UV absorbance.

Another method to be considered is Surface Plasmon Resonance (SPR). SPR can detect binding of label free molecules on the surface of a chip. If the desired protein is an antibody, binding can be translated directly to the activity of the protein. One can express the active concentration of the protein as the percent of the total protein. SPR can be a powerful method for quickly determining protein activity and overall yield. It is a powerful technology that requires an instrument to perform.

Analytical

Denaturing-condition Electrophoresis

Gel electrophoresis is a common laboratory technique that can be used both as preparative and analytical method. The principle of electrophoresis relies on the movement of a charged ion in an electric field. In practice, the proteins are denatured in a solution containing a detergent (SDS). In these conditions, the proteins are unfolded and coated with negatively charged detergent molecules. The proteins in SDS-PAGE are separated on the sole basis of their size.

In analytical methods, the protein migrate as bands based on size. Each band can be detected using stains such as Coomassie blue dye or silver stain. Preparative methods to purify large amounts of protein, require the extraction of the protein from the electrophoretic gel. This extraction may involve excision of the gel containing a band, or eluting the band directly off the gel as it runs off the end of the gel.

In the context of a purification strategy, denaturing condition electrophoresis provides an improved resolution over size exclusion chromatography, but does not scale to large quantity of proteins in a sample as well as the late chromatography columns.

Non-denaturing-condition Electrophoresis

Equipment for preparative gel electrophoresis: electrophoresis chamber, peristaltic pump, fraction collector, buffer recirculation pump and UV detector (in a refrigerator), power supply and recorder (on a table)

An important non-denaturing electrophoretic procedure for isolating bioactive metalloproteins in complex protein mixtures is quantitative native PAGE. The intactness or the structural integrity of the isolated protein has to be determined by an independent method.

References

- Petsko GA, Ringe D (2003). "Chapter 1: From sequence to structure". Protein structure and function. London: New Science. p. 27. ISBN 978-1405119221.
- Suzuki H (2015). "Chapter 7: Active Site Structure". How Enzymes Work: From Structure to Function. Boca Raton, FL: CRC Press. pp. 117–140. ISBN 978-981-4463-92-8.
- Krauss G (2003). "The Regulations of Enzyme Activity". Biochemistry of Signal Transduction and Regulation (3rd ed.). Weinheim: Wiley-VCH. pp. 89–114. ISBN 9783527605767.
- Cooper GM (2000). "Chapter 2.2: The Central Role of Enzymes as Biological Catalysts". The Cell: a Molecular Approach (2nd ed.). Washington (DC): ASM Press. ISBN 0-87893-106-6.
- Cox MM, Nelson DL (2013). "Chapter 6.2: How enzymes work". Lehninger Principles of Biochemistry (6th ed.). New York, N.Y.: W.H. Freeman. p. 195. ISBN 978-1464109621.
- McArdle WD, Katch F, Katch VL (2006). "Chapter 9: The Pulmonary System and Exercise". Essentials of Exercise Physiology (3rd ed.). Baltimore, Maryland: Lippincott Williams & Wilkins. pp. 312–3. ISBN 978-0781749916.
- Suzuki H (2015). "Chapter 8: Control of Enzyme Activity". How Enzymes Work: From Structure to Function. Boca Raton, FL: CRC Press. pp. 141–69. ISBN 978-981-4463-92-8.
- Suzuki H (2015). "Chapter 4: Effect of pH, Temperature, and High Pressure on Enzymatic Activity". How Enzymes Work: From Structure to Function. Boca Raton, FL: CRC Press. pp. 53–74. ISBN 978-981-4463-92-8.

- James WD, Elston D, Berger TG (2011). Andrews' Diseases of the Skin: Clinical Dermatology (11th ed.). London: Saunders/ Elsevier. p. 567. ISBN 978-1437703146.
- Tarté R (2008). Ingredients in Meat Products Properties, Functionality and Applications. New York: Springer. p. 177. ISBN 978-0-387-71327-4.
- Skett P, Gibson GG (2001). "Chapter 3: Induction and Inhibition of Drug Metabolism". Introduction to Drug Metabolism (3 ed.). Cheltenham, UK: Nelson Thornes Publishers. pp. 87–118. ISBN 978-0748760114.
- Farris PL (2009). "Economic Growth and Organization of the U.S. Starch Industry". In BeMiller JN, Whistler RL. Starch Chemistry and Technology (3rd ed.). London: Academic. ISBN 9780080926551.
- Boyer R (2002). "Chapter 6: Enzymes I, Reactions, Kinetics, and Inhibition". Concepts in Biochemistry (2nd ed.). New York, Chichester, Weinheim, Brisbane, Singapore, Toronto.: John Wiley & Sons, Inc. pp. 137–8. ISBN 0-470-00379-0. OCLC 51720783.
- Mount DM (2004). Bioinformatics: Sequence and Genome Analysis. 2. Cold Spring Harbor Laboratory Press. ISBN 0-87969-712-1.
- Garel, J. (1992). "Folding of large proteins: Multidomain and multisubunit proteins". In Creighton, T. Protein Folding (First ed.). New York: W.H. Freeman and Company. pp. 405–454. ISBN 0-7167-7027-X.

3

Primary Proteins Studied in Proteomics

The proteins studied in proteomics are cyanovirin-N, intrinsically disordered proteins, blood proteins and globular protein. Fundamentally disordered proteins are proteins that are not in a proper three-dimensional structure. This section has been carefully written to provide an easy understanding of the varied facets of proteins.

Cyanovirin-N

d Cyanovirin-N (CV-N) is a protein produced by the cyanobacterium *Nostoc ellipsosporum* that displays virucidal activity against several viruses, including human immunodeficiency virus (HIV). The virucidal activity of CV-N is mediated through specific high-affinity interactions with the viral surface envelope glycoproteins gp120 and gp41, as well as to high-mannose oligosaccharides found on the HIV envelope. In addition, CV-N is active against rhinoviruses, human parainfluenza virus, respiratory syncytial virus, and enteric viruses. The virucidal activity of CV-N against influenza virus is directed towards viral haemagglutinin. CV-N has a complex fold composed of a duplication of a tandem repeat of two homologous motifs comprising three-stranded beta-sheet and beta-hairpins.

Professor Julian Ma of St George's Hospital, South London, has a project in Kent, England to use genetically modified tobacco plants to produce the Cyanovirin and from this produce a cream which could be used to prevent HIV infection

Cyanovirin a protein with a highly complicated structure, it binds to sugars attached to HIV,it envelopes the protein and prevents it from binding to the mucosal cell surfaces in the Vagina and Rectum, this compound is also active against herpes viruses.

"development of cyanovirin has been exceedingly slow-paced. The chief of the NCI cyanovirin program, Michael Boyd, described it as "languishing." Apparently the NCI's production facilities, based on genetically manipulated cell cultures, have been diverted to other projects that the agency considers of higher priority. This is unfortunate: cyanovirin is of particular interest because of its relative safety. It is 10,000 times more toxic to HIV than it is to cells."

This protein may use the morpheein model of allosteric regulation.

Intrinsically Disordered Proteins

An intrinsically disordered protein (IDP) is a protein that lacks a fixed or ordered three-dimensional structure. IDPs cover a spectrum of states from fully unstructured to partially structured and include random coils, (pre-)molten globules, and large multi-domain proteins connected by flexible linkers. They constitute one of the main types of protein (alongside globular, fibrous and membrane proteins).

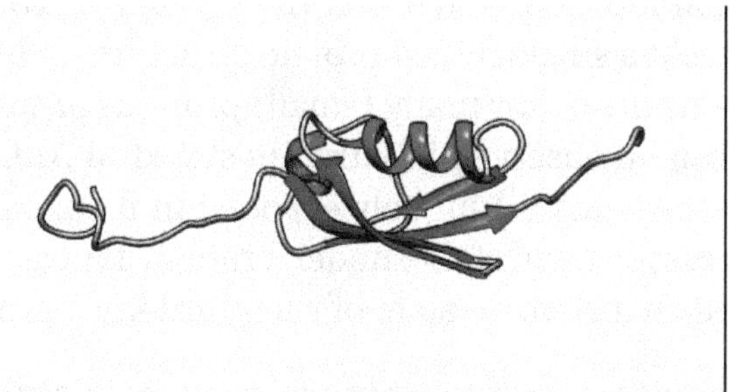

Conformational flexibility in SUMO-1 protein (PDB:1a5r). The central part shows relatively ordered structure. Conversely, the N- and C-terminal regions (left and right, respectively) show 'intrinsic disorder', although a short helical region persists in the N-terminal tail. Ten alternative NMR models were morphed. Secondary structure elements: α-helices (red), β-strands (blue arrows).

The discovery of IDPs has challenged the traditional protein structure paradigm, that protein function depends on a fixed three-dimensional structure. This dogma has been challenged over the last decades by increasing evidence from various branches of structural biology, suggesting that protein dynamics may be highly relevant for such systems. Despite their lack of stable structure, IDPs are a very large and functionally important class of proteins. In some cases, IDPs can adopt a fixed three-dimensional structure after binding to other macromolecules. Overall, IDPs are different from structured proteins in many ways and tend to have distinct properties in terms of function, structure, sequence, interactions, evolution and regulation.

History

An ensemble of NMR structures of the Thylakoid soluble phosphoprotein TSP9, which shows a largely flexible protein chain.

In the 1930s -1950s, the first protein structures were solved by protein crystallography. These early structures suggested that a fixed three-dimensional structure might be generally required to mediate biological functions of proteins. When stating that proteins have just one uniquely defined configuration, Mirsky and Pauling did not recognize that Fisher's work would have supported their thesis with his 'Lock and Key' model (1894). These publications solidified the central dogma of molecular biology in that the sequence determines the structure which, in turn, determines the function of proteins. In 1950, Karush wrote about 'Configurational Adaptability' contradicting all the assumptions and research in the 19th century. He was convinced that proteins have more than one configuration at the same energy level and can choose one when binding to other substrates. In the 1960s, Levinthal's paradox suggested that the systematic conformational search of a long polypeptide is unlikely to yield a single folded protein structure on biologically relevant timescales (i.e. seconds to minutes). Curiously, for many (small) proteins or protein domains, relatively rapid and efficient refolding can be observed in vitro. As stated in Anfinsen's Dogma from 1973, the fixed 3D structure of these proteins is uniquely encoded in its primary structure (the amino acid sequence), is kinetically accessible and stable under a range of (near) physiological conditions, and can therefore be considered as the native state of such "ordered" proteins.

During the subsequent decades, however, many large protein regions could not be assigned in x-ray datasets, indicating that they occupy multiple positions which average out in electron density maps. The lack of a fixed, unique positions relative to the crystal lattice suggested that these regions were "disordered". Additional techniques for determining protein structures, such as NMR, demonstrated the presence of large flexible linkers and termini in many solved structural ensembles. It is now generally accepted that proteins exist as an ensemble of similar structures with some regions more constrained than others. Intrinsically Unstructured Proteins (IUPs) occupy the extreme end of this spectrum of flexibility, whereas IDPs also include proteins of considerable local structure tendency or flexible multidomain assemblies.These highly dynamic disordered regions of proteins have subsequently been linked to functionally important phenomena such as allosteric regulation and enzyme catalysis.

In the 2000s, bioinformatic predictions of intrinsic disorder in proteins indicated that intrinsic disorder is more common in sequenced/predicted proteomes than in known structures in the protein database. Based on DISOPRED2 prediction, long (>30 residue) disordered segments occur in 2.0% of archaean, 4.2% of eubacterial and 33.0% of eukaryotic proteins. In 2001, Dunker published his paper 'Intrinsically Disordered Proteins' questioning whether the newly found information was ignored for 50 years.

In the 2010s it became clear that IDPs are highly abundant among disease-related proteins.

Biological Roles

Many disordered proteins have the binding affinity with their receptors regulated by post-translational modification, thus it has been proposed that the flexibility of disordered proteins facilitates the different conformational requirements for binding the modifying enzymes as well as their receptors. Intrinsic disorder is particularly enriched in proteins implicated in cell signaling, transcription and chromatin remodeling functions.

Flexible Linkers

Disordered regions are often found as flexible linkers (or loops) connecting two globular or transmembrane domains. Linker sequences vary greatly in length and amino acid sequence, but are similar in amino acid composition (rich in polar uncharged amino acids). Flexible linkers allow the connecting domains to freely twist and rotate through space to recruit their binding partners or for those binding partners to induce larger scale interdomain conformation changes via protein domain dynamics.

Linear Motifs

Linear motifs are short disordered segments of proteins that mediate functional interactions with other proteins or other biomolecules (RNA, DNA, sugars etc.). Many roles or linear motifs are associated with cell regulation, for instance in control of cell shape, subcellular localisation of individual proteins and regulated protein turnover. Often, post-translational modifications such as phosphorylation tune the affinity (not rarely by several orders of magnitude) of individual linear motifs for specific interactions. Relatively rapid evolution and a relatively small number of structural restraints for establishing novel (low-affinity) interfaces make it particularly challenging to detect linear motifs but their widespread biological roles and the fact that many viruses mimick/hijack linear motifs to efficiently recode infected cells underlines the timely urgency of research on this very challenging and exciting topic. Unlike globular proteins IDPs do not have spatially-disposed active pockets. Nevertheless, in 80% of IDPs (~3 dozens) subjected to detailed structural characterization by NMR there are linear motifs termed PreSMos (pre-structured motifs) that are transient secondary structural elements primed for target recognition. In several cases it has been demonstrated that these transient structures become full and stable secondary structures, e.g., helices, upon target binding. Hence, PreSMos are the putative active sites in IDPs.

Coupled Folding and Binding

Many unstructured proteins undergo transitions to more ordered states upon binding to their targets (e.g. Molecular Recognition Features (MoRFs)). The coupled folding and binding may be local, involving only a few interacting residues, or it might involve an entire protein domain. It was recently shown that the coupled folding and binding allows the burial of a large surface area that would be possible only for fully structured proteins if they were much larger. Moreover, certain disordered regions might serve as "molecular switches" in regulating certain biological function by switching to ordered conformation upon molecular recognition like small molecule-binding, DNA/RNA binding, ion interactions etc.

The ability of disordered proteins to bind, and thus to exert a function, shows that stability is not a required condition. Many short functional sites, for example Short Linear Motifs are over-represented in disordered proteins.

Disorder in the Bound State (Fuzzy Complexes)

Intrinsically disordered proteins can retain their conformational freedom even when they bind specifically to other proteins. The structural disorder in bound state can be static or dynamic. In fuzzy complexes structural multiplicity is required for function and the manipulation of the bound disordered region changes activity. The conformational ensemble of the complex is modulated via post-translational modifications or protein interactions. Specificity of DNA binding proteins often depends on the length of fuzzy regions, which is varied by alternative splicing.

Structural Aspects

Intrinsically disordered proteins adapt many different structures in vivo according to the cell's conditions, creating a structural or conformational ensemble.

Therefore their structures are strongly function-related. However, only few proteins are fully disordered in their native state. Disorder is mostly found in intrinsically disordered regions (IDRs) within an otherwise well-structured protein. The term intrinsically disordered protein (IDP) therefore includes proteins that contain IDRs as well as fully disordered proteins.

The existence and kind of protein disorder is encoded in its amino acid sequence. In general, IDPs are characterized by a low content of bulky hydrophobic amino acids and a high proportion of polar and charged amino acids, usually referred to as low hydrophobicity. This property leads to good interactions with water. Furthermore high net charges promote disorder because of electrostatic repulsion resulting from equally charged residues. Thus disordered sequences cannot sufficiently bury a hydrophobic core to fold into stable globular proteins. In some cases, hydrophobic clusters in disordered sequences provide the clues for identifying the regions that undergo coupled folding and binding (refer to biological roles). Many disordered proteins reveal regions without any regular secondary structure These regions can be termed as flexible, compared to structured loops. While the latter are rigid and contain only one set of Ramachandran angles, IDPs involve multiple sets of angles. The term flexibility is also used for well-structured proteins, but describes a different phenomenon in the context of disordered proteins. Flexibility in structured proteins is bound to an equilibrium state, while it is not so in IDPs. Many disordered proteins also reveal low complexity sequences, i.e. sequences with over-representation of a few residues. While low complexity sequences are a strong indication of disorder, the reverse is not necessarily true, that is, not all disordered proteins have low complexity sequences. Disordered proteins have a low content of predicted secondary structure.

Experimental Validation

Intrinsically unfolded proteins, once purified, can be identified by various experimental methods. The primary method to obtain information on disordered regions of a protein is NMR spectroscopy. The lack of electron density in X-ray crystallographic studies may also be a sign of disorder.

Folded proteins have a high density (partial specific volume of 0.72-0.74 mL/g) and commensurately small radius of gyration. Hence, unfolded proteins can be detected by methods that are sensitive to molecular size, density or hydrodynamic drag, such as size exclusion chromatography, analytical ultracentrifugation, Small angle X-ray scattering (SAXS), and measurements of the diffusion constant. Unfolded proteins are also characterized by their lack of secondary structure, as assessed by far-UV (170-250 nm) circular dichroism (esp. a pronounced minimum at ~200 nm) or infrared spectroscopy. Unfolded proteins also have exposed backbone peptide groups exposed to solvent, so that they are readily cleaved by proteases, undergo rapid hydrogen-deuterium exchange and exhibit a small dispersion (<1 ppm) in their 1H amide chemical shifts as measured by NMR. (Folded proteins typically show dispersions as large as 5 ppm for the amide protons.) Recently, new methods including Fast parallel proteolysis (FASTpp) have been introduced, which allow to determine the fraction folded/disordered without the need for purification. Even subtle differences in the stability of missense mutations, protein partner binding and (self)polymerisation-induced folding of (e.g.) coiled-coils can be detected using FASTpp as recently demonstrated using the tropomyosin-troponin protein interaction. Fully unstructured protein regions can be experimentally validated by their hypersusceptibility to proteolysis using short digestion times and low protease concentrations.

Bulk methods to study IDP structure and dynamics include SAXS for ensemble shape information, NMR for atomistic ensemble refinement, Fluorescence for visualising molecular interactions and conformational transitions, x-ray crystallography to highlight more mobile regions in otherwise rigid protein crystals, cryo-EM to reveal less fixed parts of proteins, light scattering to monitor size distributions of IDPs or their aggregation kinetics, Circular Dichroism to monitor secondary structure of IDPs.

Single-molecule methods to study IDPs include spFRET to study conformational flexibility of IDPs and the kinetics of structural transitions, optical tweezers for high-resolution insights into the ensembles of IDPs and their oligomers or aggregates, nanopores to reveal global shape distributions of IDPs, magnetic tweezers to study structural transitions for long times at low forces, high-speed AFM to visualise the spatio-temporal flexibility of IDPs directly.

Disorder Prediction

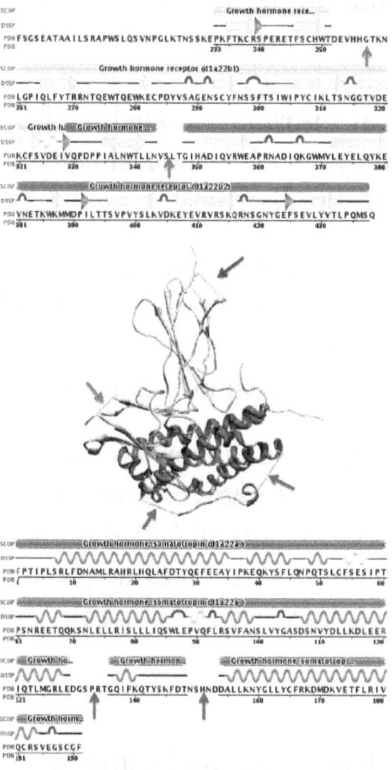

REMARK465 - missing electron densities in X-ray structure representing protein disorder (PDB: 1a22, human growth hormone bound to receptor). Compilation of screenshots from PDB database and molecule representation via VMD. Blue and red arrows point to missing residues on receptor and growth hormone, respectively.

Disorder prediction algorithms can predict Intrinsic Disorder (ID) propensity with high accuracy (approaching around 80%) based on primary sequence composition, similarity to unassigned segments in protein x-ray datasets, flexible regions in NMR studies and physico-chemical properties of amino acids.

Distinguishing IDPs from Well-structured Proteins

Separating disordered from ordered proteins is essential for disorder prediction. One of the first steps to find a factor that distinguishes IDPs from non-IDPs is to specify biases within the amino acid composition. The following hydrophilic, charged amino acids A, R, G, Q, S, P, E and K have been characterized as disorder-promoting amino acids, while order-promoting amino acids W, C, F, I, Y, V, L, and N are hydrophobic and uncharged. The remaining amino acids H, M, T and D are ambiguous, found in both ordered and unstructured regions. This information is the basis of most sequence-based predictors. Regions with little to no secondary structure, also known as NORS (NO Regular Secondary structure) regions, and low-complexity regions can easily be detected. However, not all disordered proteins contain such low complexity sequences.

Prediction Methods

Determining disordered regions from biochemical methods is very costly and time-consuming. Due to the variable nature of IDPs, only certain aspects of their structure can be detected, so that a full characterization requires a large number of different methods and experiments. This further increases the expense of IDP determination. In order to overcome this obstacle, computer-based methods are created for predicting protein structure and function. It is one of the main goals of bioinformatics to derive knowledge by prediction. Predictors for IDP function are also being developed, but mainly use structural information such as linear motif sites. There are different approaches for predicting IDP structure, such as neural networks or matrix calculations, based on different structural and/or biophysical properties.

Many computational methods exploit sequence information to predict whether a protein is disordered. Notable examples of such software include IUPRED and Disopred. Different methods may use different definitions of disorder. Meta-predictors show a new concept, combining different primary predictors to create a more competent and exact predictor.

Due to the different approaches of predicting disordered proteins, estimating their relative accuracy is fairly difficult. For example, neural networks are often trained on different datasets. The disorder prediction category is a part of biannual CASP experiment that is designed to test methods according accuracy in finding regions with missing 3D structure (marked in PDB files as REMARK465, missing electron densities in X-ray structures).

Disorder and Disease

Intrinsically unstructured proteins have been implicated in a number of diseases. Aggregation of misfolded proteins is the cause of many synucleinopathies and toxicity as those proteins start binding to each other randomly and can lead to cancer or cardiovascular diseases. Thereby, misfolding can happen spontaneously because millions of copies of proteins are made during the lifetime of an organism. The aggregation of the intrinsically unstructured protein α-Synuclein is thought to be responsible. The structural flexibility of this protein together with its susceptibility to modification in the cell leads to misfolding and aggregation. Genetics, oxidative and nitrative stress as well as mitochondrial impairment impact the structural flexibility of the unstructured α-Synuclein protein and associated disease mechanisms. Many key oncogenes have large intrinsically unstructured regions, for example p53 and BRCA1. These regions of the proteins are responsible for mediating many of their interactions. Taking the cell's native defense mechanisms as a model drugs can be developed, trying to block the place of noxious substrates and inhibiting them, and thus counteracting the disease.

Computer Simulations

Structural and dynamical properties of intrinsically unstructured proteins are being studied by molecular dynamics simulations. Findings from these simulations suggest a highly flexible conformational ensemble of intrinsically disordered proteins at different temperatures which is related to the presence of low free energy barriers.

Effects of confinement have also recently been addressed. For example, these studies suggest that confinement tends to increase the population of turn structures with respect to the population of coils and β-hairpins.

Moreover, various protocols and methods of analyzing IDPs, such as studies based on quantitative analysis of GC content in genes and their respective chromosomal bands, have been used to understand functional IDP segments.

Pioneering IDP Research Labs

In the last ten years, a great number of laboratories have investigated protein disorders using both experimental (e.g. SAXS-NMR, single-molecule fluorescence) and computational (analysis of protein structure) techniques.

Blood Proteins

Blood proteins, also termed plasma proteins or serum proteins, are proteins present in blood plasma. They serve many different functions, including transport of lipids, hormones, vitamins and minerals in the circulatory system and the regulation of acellular activity and functioning of the immune system. Other blood proteins act as enzymes, complement components, protease inhibitors or kinin precursors. Contrary to popular belief, hemoglobin is not a blood protein, as it is carried within red blood cells, rather than in the blood serum.

Serum albumin accounts for 55% of blood proteins, and is a major contributor to maintaining the osmotic pressure of plasma to assist in the transport of lipids and steroid hormones. Globulins make up 38% of blood proteins and transport ions, hormones, and lipids assisting in immune function. Fibrinogen comprises 7% of blood proteins; conversion of fibrinogen to insoluble fibrin is essential for blood clotting. The remainder of the plasma proteins (1%) are regulatory proteins, such as enzymes, proenzymes, and hormones. All blood proteins are synthesized in liver except for the gamma globulins.

Separating serum proteins by electrophoresis is a valuable diagnostic tool as well as a way to monitor clinical progress. Current research regarding blood plasma proteins is centered on performing proteomics analyses of serum/plasma in the search for biomarkers. These efforts started with two-dimensional gel electrophoresis efforts in the 1970s and in more recent times this research has been performed using LC-tandem MS based proteomics. The normal laboratory value of serum total protein is around 7 g/dL.

Families of blood proteins:

Blood protein	Normal level	%	Function
Albumins	3.5-5.0 g/dl	55%	maintain colloid osmotic pressure; create oncotic pressure and transport insoluble molecules
Globulins	2.0-2.5 g/dl	38%	participate in immune system
Fibrinogen	0.2-0.45 g/dl	7%	Blood coagulation
Regulatory proteins		<1%	Regulation of gene expression
Clotting factors		<1%	Conversion of fibrinogen into fibrin

Examples of specific blood proteins:

- Prealbumin (transthyretin)
- Alpha 1 antitrypsin (neutralizes trypsin that has leaked from the digestive system)
- Alpha 1 acid glycoprotein
- Alpha 1 fetoprotein
- alpha2-macroglobulin
- Gamma globulins
- Beta-2 microglobulin
- Haptoglobin
- Ceruloplasmin
- Complement component 3
- Complement component 4
- C-reactive protein (CRP)
- Lipoproteins (chylomicrons, VLDL, LDL, HDL)
- Transferrin
- Prothrombin
- MBL or MBP

Globular Protein

Globular proteins or spheroproteins are spherical ("globe-like") proteins and are one of the common protein types (the others being fibrous, disordered and membrane proteins). Globular proteins are somewhat water-soluble (forming colloids in water), unlike the fibrous or membrane proteins. There are multiple fold classes of globular proteins, since there are many different architectures that can fold into a roughly spherical shape.

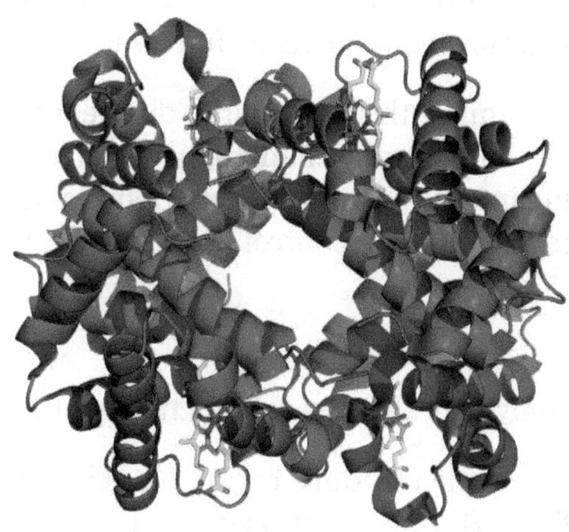

3-dimensional structure of hemoglobin, a globular protein.

The term globin can refer more specifically to proteins including the globin fold.

Globular Structure and Solubility

The term globular protein is quite old (dating probably from the 19th century) and is now somewhat archaic given the hundreds of thousands of proteins and more elegant and descriptive structural motif vocabulary. The globular nature of these proteins can be determined without the means of modern techniques, but only by using ultracentrifuges or dynamic light scattering techniques.

The spherical structure is induced by the protein's tertiary structure. The molecule's apolar (hydrophobic) amino acids are bounded towards the molecule's interior whereas polar (hydrophilic) amino acids are bound outwards, allowing dipole-dipole interactions with the solvent, which explains the molecule's solubility.

Globular proteins are only marginally stable because the free energy released when the protein folded into its native conformation is relatively small. This is because protein folding requires entropic cost. As a primary sequence of a polypeptide chain can form numerous conformations, native globular structure restricts its conformation to a few only. It results in a decrease in randomness, although non-covalent interactions such as hydrophobic interactions stabilize the structure.

Although it is still unknown how proteins fold up naturally, new evidence has helped advance understanding. Part of the protein folding problem is that several non-covalent, weak interactions are formed, such as hydrogen bonds and Van der Waals interactions. Via several techniques, the mechanism of protein folding is currently being studied. Even in the protein's denatured state, it can be folded into the correct structure.

Globular proteins seem to have two mechanisms for protein folding, either the diffusion-collision model or nucleation condensation model, although recent findings have shown globular proteins, such as PTP-BL PDZ2, that fold with characteristic features of both models. These new findings have shown that the transition states of proteins may affect the way they fold. The folding of globular proteins has also recently been connected to treatment of diseases, and anti-cancer ligands have been developed which bind to the folded but not the natural protein. These studies have shown that the folding of globular proteins affects its function.

By the second law of thermodynamics, the free energy difference between unfolded and folded states is contributed by enthalpy and entropy changes. As the free energy difference in a globular protein that results from folding into its native conformation is small, it is marginally stable, thus providing a rapid turnover rate and effective control of protein degradation and synthesis.

Role

Unlike fibrous proteins which only play a structural function, globular proteins can act as:

- Enzymes, by catalyzing organic reactions taking place in the organism in mild conditions and with a great specificity. Different esterases fulfill this role.
- Messengers, by transmitting messages to regulate biological processes. This function is done by hormones, i.e. insulin etc.
- Transporters of other molecules through membranes
- Stocks of amino acids.
- Regulatory roles are also performed by globular proteins rather than fibrous proteins.
- Structural proteins, e.g., actin and tubulin, which are globular and soluble as monomers, but polymerize to form long, stiff fibers

Members

Among the most known globular proteins is hemoglobin, a member of the globin protein family. Other globular proteins are the immunoglobulins (IgA, IgD, IgE, IgG and IgM), and alpha, beta and gamma globulins. Nearly all enzymes with major metabolic functions are globular in shape, as well as many signal transduction proteins.

Albumins are also globular proteins, although, unlike all of the other globular proteins, they are completely soluble in water.

4

Various Interactive Protein Domains

The various interactive proteins domains discussed are SH2 domain, SH3 domain, Phosphotyrosine-binding domain, LIM domain, FERM domain, RNA-binding protein and pleckstrin homology domain. If the domain size of a protein is about 60 amino acids, it's known as SRC homology 3 domain. The diverse interactive protein domains have been carefully explained in this chapter.

SH2 Domain

The SH2 (Src Homology 2) domain is a structurally conserved protein domain contained within the Src oncoprotein and in many other intracellular signal-transducing proteins. SH2 domains allow proteins containing those domains to dock to phosphorylated tyrosine residues on other proteins. SH2 domains are commonly found in adapter proteins that aid in the signal transduction of receptor tyrosine kinase pathways.

Introduction

Protein-protein interactions play a major role in cellular growth and development. Modular domains, which are the subunits of a protein, moderate these protein interactions by identifying short peptide sequences. These peptide sequences determine the binding partners of each protein. One of the more prominent domains is the SH2 domain. SH2 domains play a vital role in cellular communication. Its length is approximately 100 amino acids long and it is found within 111 human proteins. Regarding its structure, it contains 2 alpha helices and 7 beta strands. Research has shown that it has a high affinity to phosphorylated tyrosine residues and it is known to identify a sequence of 3-6 amino acids within a peptide motif.

Binding and Phosphorylation

SH2 domains typically bind a phosphorylated tyrosine residue in the context of a longer peptide motif within a target protein, and SH2 domains represent the largest class of known pTyr-recognition domains.

Phosphorylation of tyrosine residues in a protein occurs during signal transduction and is carried out by tyrosine kinases. In this way, phosphorylation of a substrate by tyrosine kinases acts as a switch to trigger binding to an SH2 domain-containing protein. Many tyrosine containing short linear motifs that bind to SH2 domains are conserved across a wide variety of higher Eukaryotes. The intimate relationship between tyrosine kinases and SH2 domains is supported by their coordinate emergence during eukaryotic evolution.

Diversity

SH2 domains are not present in yeast and appear at the boundary between protozoa and animalia in organisms such as the social amoeba Dictyostelium discoideum.

A detailed bioinformatic examination of SH2 domains of human and mouse reveals 120 SH2 domains contained within 115 proteins encoded by the human genome, representing a rapid rate of evolutionary expansion among the SH2 domains.

A large number of SH2 domain structures have been solved and many SH2 proteins have been knocked out in mice. Information generated on the Mouse Knockouts can be found on the sh2.uchicago.edu website.

Function

The function of SH2 domains is to specifically recognize the phosphorylated state of tyrosine residues, thereby allowing SH2 domain-containing proteins to localize to tyrosine-phosphorylated sites. This process constitutes the fundamental event of signal transduction through a membrane, in which a signal in the extracellular compartment is "sensed" by a receptor and is converted in the intracellular compartment to a different chemical form, i.e. that of a phosphorylated tyrosine. Tyrosine phosphorylation leads to activation of a cascade of protein-protein interactions whereby SH2 domain-containing proteins are recruited to tyrosine-phosphorylated sites. This process initiates a series of events which eventually result in altered patterns of gene expression or other cellular responses. The SH2 domain, which was first identified in the oncoproteins Src and Fps, is about 100 amino-acid residues long. It functions as a regulatory module of intracellular signaling cascades by interacting with high affinity to phosphotyrosine-containing target peptides in a sequence-specific and strictly phosphorylation-dependent manner.

Examples

Human proteins containing this domain include:

- ABL1; ABL2
- BCAR3; BLK; BLNK; BMX; BTK
- CHN2; CISH; CRK; CRKL; CSK
- DAPP1
- FER; FES; FGR; FRK; FYN
- GRAP; GRAP2; GRB10; GRB14; GRB2; GRB7
- HCK; HSH2D
- INPP5D; INPPL1; ITK; JAK2; LCK; LCP2; LYN
- MATK; NCK1; NCK2

- PIK3R1; PIK3R2; PIK3R3; PLCG1; PLCG2; PTK6; PTPN11; PTPN6; RASA1
- SH2B1; SH2B2; SH2B3; SH2D1A; SH2D1B; SH2D2A; SH2D3A; SH2D3C; SH2D4A; SH2D4B; SH2D5; SH2D6; SH3BP2; SHB; SHC1; SHC3; SHC4; SHD; SHE
- SLA; SLA2
- SOCS1; SOCS2; SOCS3; SOCS4; SOCS5; SOCS6; SOCS7
- SRC; SRMS
- STAT1; STAT2; STAT3; STAT4; STAT5A; STAT5B; STAT6
- SUPT6H; SYK
- TEC; TENC1; TNS; TNS1; TNS3; TNS4; TXK
- VAV1; VAV2; VAV3
- YES1; ZAP70

SH3 Domain

The SRC Homology 3 Domain (or SH3 domain) is a small protein domain of about 60 amino acids residues. Initially, SH3 was described as a conserved sequence in the viral adaptor protein v-Crk. This domain is also present in the molecules of phospholipase and several cytoplasmic tyrosine kinases such as Abl and Src. It has also been identified in several other protein families such as: PI3 Kinase, Ras GTPase-activating protein, CDC24 and cdc25. SH3 domains are found in proteins of signaling pathways regulating the cytoskeleton, the Ras protein, and the Src kinase and many others. The SH3 proteins interact with adaptor proteins and tyrosine kinases. Interacting with tyrosine kinases SH3 proteins usually bind far away from the active site. Approximately 300 SH3 domains are found in proteins encoded in the human genome.

Structure

The SH3 domain has a characteristic beta-barrel fold that consists of five or six β-strands arranged as two tightly packed anti-parallel β sheets. The linker regions may contain short helices. The SH3-type fold is an ancient fold found in eukaryotes as well as prokaryotes.

Peptide Binding

The classical SH3 domain is usually found in proteins that interact with other proteins and mediate assembly of specific protein complexes, typically via binding to proline-rich peptides in their respective binding partner. Classical SH3 domains are restricted in humans to intracellular proteins, although the small human MIA family of extracellular proteins also contain a domain with an SH3-like fold.

Many SH3-binding epitopes of proteins have a consensus sequence that can be represented as a regular expression or Short linear motif:

-X-P-p-X-P-

1 2 3 4 5

with 1 and 4 being aliphatic amino acids, 2 and 5 always and 3 sometimes being proline. The sequence binds to the hydrophobic pocket of the SH3 domain. More recently, SH3 domains that bind to a core consensus motif R-x-x-K have been described. Examples are the C-terminal SH3 domains of adaptor proteins like Grb2 and Mona (a.k.a. Gads, Grap2, Grf40, GrpL etc.). Other SH3 binding motifs have emerged and are still emerging in the course of various molecular studies, highlighting the versatility of this domain.

Proteins With SH3 Domain

- Adaptor proteins
- CDC24
- Cdc25
- PI3 kinase
- Phospholipase
- Ras GTPase-activating protein
- Vav proto-oncogene
- GRB2
- p54 S6 kinase 2 (S6K2)
- SH3D21
- C10orf76 (potentially)
- STAC3
- Some myosins
- SHANK1,2,3
- ARHGAP12
- C8orf46
- TANGO1
- Integrase
- Focal Adhesion Kinase (FAK, PTK2)
- Proline-rich tyrosine kinase (Pyk2, CADTK, PTK2beta)

Phosphotyrosine-binding Domain

In molecular biology, Phosphotyrosine-binding domains are protein domains which bind to phosphotyrosine.

The phosphotyrosine-binding domain (PTB, also phosphotyrosine-interaction or PI domain) in the protein tensin tends to be found at the C-terminus. Tensin is a multi-domain protein that binds to actin filaments and functions as a focal-adhesion molecule (focal adhesions are regions of plasma membrane through which cells attach to the extracellular matrix). Human tensin has actin-binding sites, an SH2 (*Pfam PF00017*) domain and a region similar to the tumour suppressor PTEN. The PTB domain interacts with the cytoplasmic tails of beta integrin by binding to an NPXY motif.

The phosphotyrosine-binding domain of insulin receptor substrate-1 is not related to the phosphotyrosine-binding domain of tensin. Insulin receptor substrate-1 proteins contain both a pleckstrin homology domain and a phosphotyrosine binding (PTB) domain. The PTB domains facilitate interaction with the activated tyrosine-phosphorylated insulin receptor. The PTB domain is situated towards the N terminus. Two arginines in this domain are responsible for hydrogen bonding phosphotyrosine residues on an Ac-LYASSNPApY-NH2 peptide in the juxtamembrane region of the insulin receptor. Further interactions via "bridged" water molecules are coordinated by residues an Asn and a Ser residue. The PTB domain has a compact, 7-stranded beta-sandwich structure, capped by a C-terminal helix. The substrate peptide fits into an L-shaped surface cleft formed from the C-terminal helix and strands 5 and 6.

Human Proteins Containing These Domains

APBA1; APBA2; APBA3; EPS8; EPS8L1; EPS8L2; EPS8L3; TENC1; TNS; TNS1; TNS3; TNS4; DOK1; DOK2; DOK3; DOK4; DOK5; DOK6; DOK7; FRS2; FRS3; IRS1; IRS2; IRS4; TLN1; TLN2

LIM Domain

LIM domains are protein structural domains, composed of two contiguous zinc finger domains, separated by a two-amino acid residue hydrophobic linker. They are named after their initial discovery in the proteins Lin11, Isl-1 & Mec-3. LIM-domain containing proteins have been shown to play roles in cytoskeletal organisation, organ development and oncogenesis. LIM-domains mediate protein–protein interactions that are critical to cellular processes.

LIM domains have highly divergent sequences, apart from certain key residues. The sequence divergence allow a great many different binding sites to be grafted onto the same basic domain. The conserved residues are those involved in zinc binding or the hydrophobic core of the protein. The sequence signature of LIM domains is as follows:

$[C]\text{-}[X]_{2-4}\text{-}[C]\text{-}[X]_{13-19}\text{-}[W]\text{-}[H]\text{-}[X]_{2-4}\text{-}[C]\text{-}[F]\text{-}[LVI]\text{-}[C]\text{-}[X]_{2-4}\text{-}[C]\text{-}[X]_{13-20}\text{-}C\text{-}[X]_{2-4}\text{-}[C]$

LIM domains frequently occur in multiples, as seen in proteins such as TES, LMO4, and can also be attached to other domains in order to confer a binding or targeting function upon them, such as LIM-kinase.

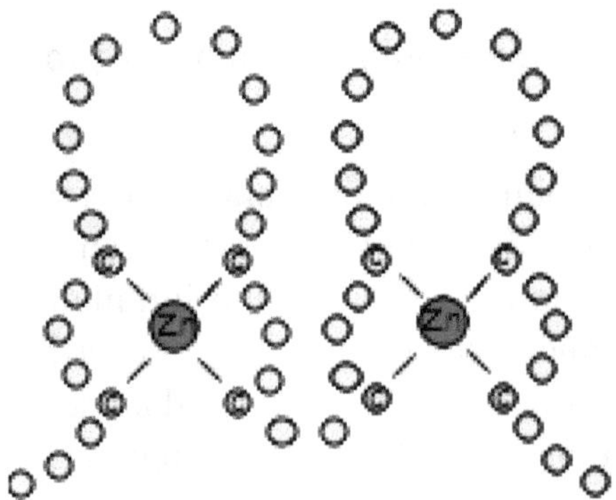

LIM domain organisation

The LIM superclass of genes have been classified into 14 classes: ABLIM, CRP, ENIGMA, EPLIN, LASP, LHX, LMO, LIMK, LMO7, MICAL, PXN, PINCH, TES, and ZYX. Six of these classes (i.e., ABLIM, MICAL, ENIGMA, ZYX, LHX, LMO7) originated in the stem lineage of animals, and this expansion is thought to have made a major contribution to the origin of animal multicellularity.

LIM domains are also found in various bacterial lineages where they are typically fused to a metallopeptidase domain. Some versions show fusions to an inactive P-loop NTPase at their N-terminus and a single transmembrane helix. These domain fusions suggest that the prokaryotic LIM domains are likely to regulate protein processing at the cell membrane. The domain architectural syntax is remarkable parallel to those of the prokaryotic versions of the B-box zinc finger and the AN1 zinc finger domains.

Sterile Alpha Motif

In molecular biology, the protein domain Sterile alpha motif (or SAM) is a putative protein interaction module present in a wide variety of proteins involved in many biological processes. The SAM domain that spreads over around 70 residues is found in diverse eukaryotic organisms. SAM domains have been shown to homo- and hetero-oligomerise, forming multiple self-association architectures and also binding to various non-SAM domain-containing proteins, nevertheless with a low affinity constant.

SAM domains also appear to possess the ability to bind RNA. Smaug a protein that helps to establish a morphogen gradient in Drosophila embryos by repressing the translation of nanos (nos) mRNA binds to the 3' untranslated region (UTR) of nos mRNA via two similar hairpin structures. The 3D crystal structure of the Smaug RNA-binding region shows a cluster of positively charged residues on the Smaug-SAM domain, which could be the RNA-binding surface. This electropositive potential is unique among all previously determined SAM-domain structures and is conserved among Smaug-SAM homologs. These results suggest that the SAM domain might have a primary role in RNA binding.

Structural analyses show that the SAM domain is arranged in a small five-helix bundle with two large interfaces. In the case of the SAM domain of EPHB2, each of these interfaces is able to form dimers. The presence of these two distinct intermonomers binding surface suggest that SAM could form extended polymeric structures.

Fungal SAM

In molecular biology, the protein domain Ste50p mainly in fungi and some other types of eukaryotes. It plays a role in the mitogen-activated protein kinase cascades, a type of cell signalling that helps the cell respond to external stimuli, more specifically mating, cell growth, and osmo-tolerance in fungi.

Function

The protein domain Ste50p has a role in detecting pheromones for mating. It is thought to be found bound to Ste11p in order to prolong the pheromone-induced signaling response. Furthermore, it is also involved in aiding the cell to respond to nitrogen starvation.

Structure

The fungal Ste50p SAM consists of six helices, which form a compact, globular fold. It is a monomer in solution and often undergoes heterodimerisation (and in some cases oligomerisation) of the protein.

Protein Interaction

The SAM domain of Ste50p often interacts with the SAM domain of Ste11p. They form bonds through this association. It is important to note that the SAM domain of one protein will bind to the SAM of a different protein. SAM domains do not self-associate in vitro. There is significant evidence for Ste50p oligomerization in vivo.

Human Proteins Containing this Domain

ANKS1A; ANKS1B; ANKS3; ANKS4B; ANKS6; BFAR; BICC1; CASKIN1; CASKIN2; CENTD1; CNKSR2; CNKSR3; DDHD2; EPHA1; EPHA10; EPHA2; EPHA5; EPHA6; EPHA7; EPHA8; EPHB1; EPHB2; EPHB3; EPHB4; FAM59A; HPH2; INPPL1; L3MBTL3; PHC1; PHC2; PHC3; PPFIA1; PPFIA2; PPFIA3; PPFIA4; PPFIBP1; PPFIBP2; SAMD1; SAMD13; SAMD14; SAMD3; SAMD4A; SAMD4B; SAMD5; SAMD7; SAMD8; SAMD9; SCMH1; SCML1; SCML2; SEC23IP; SGMS1; SHANK1; SHANK2; SHANK3; STARD13; UBP1; USH1G; ZCCHC14; p63; p73;

FERM Domain

In molecular biology, the FERM domain (F for 4.1 protein, E for ezrin, R for radixin and M for moesin) is a widespread protein module involved in localising proteins to the plasma membrane. FERM domains are found in a number of cytoskeletal-associated proteins that associate with various proteins at the interface between the plasma membrane and the cytoskeleton. The FERM domain is located at the N terminus in the majority of proteins in which it is found.

Structure and Function

Ezrin, moesin, and radixin are highly related proteins (ERM protein family), but the other proteins in which the FERM domain is found do not share any region of similarity outside of this domain. ERM proteins are made of three domains, the FERM domain, a central helical domain and a C-terminal tail domain, which binds F-actin. The amino-acid sequence of the FERM domain is highly conserved among ERM proteins and is responsible for membrane association by direct binding to the cytoplasmic domain or tail of integral membrane proteins. ERM proteins are regulated by an intramolecular association of the FERM and C-terminal tail domains that masks their binding sites for other molecules. For cytoskeleton-membrane cross-linking, the dormant molecules becomes activated and the FERM domain attaches to the membrane by binding specific membrane proteins, while the last 34 residues of the tail bind actin filaments. Aside from binding to membranes, the activated FERM domain of ERM proteins can also bind the guanine nucleotide dissociation inhibitor of Rho GTPase (RhoDGI), which suggests that in addition to functioning as a cross-linker, ERM proteins may influence Rho signalling pathways. The crystal structure of the FERM domain reveals that it is composed of three structural modules (F1, F2, and F3) that together form a compact clover-shaped structure. The N-terminal module is ubiquitin-like. The C-terminal module is a PH-like domain.

The FERM domain has also been called the amino-terminal domain, the 30kDa domain, 4.1N30, the membrane-cytoskeletal-linking domain, the ERM-like domain, the ezrin-like domain of the band 4.1 superfamily, the conserved N-terminal region, and the membrane attachment domain.

Examples

FERM domain containing proteins include:

- Band 4.1, which links the spectrin-actin cytoskeleton of erythrocytes to the plasma membrane.
- Ezrin, a component of the undercoat of the microvilli plasma membrane.
- Moesin, which is probably involved in binding major cytoskeletal structures to the plasma membrane.
- Radixin, which is involved in the binding of the barbed end of actin filaments to the plasma membrane in the undercoat of the cell-to-cell Adherens junction.
- Talin, a cytoskeletal protein concentrated in regions of cell-substratum contact and, in lymphocytes, of cell-cell contacts.
- Filopodin, a slime mould protein that binds actin and which is involved in the control of cell motility and chemotaxis.
- Merlin (or schwannomin).
- Protein NBL4.
- Unconventional myosins X, VIIa and XV, which are mutated in congenital deafness.

- Focal-adhesion kinases (FAKs), cytoplasmic protein tyrosine kinases involved in signalling through integrins.
- Janus tyrosine kinases (JAKs), cytoplasmic tyrosine kinases that are non-covalently associated with the cytoplasmic tails of receptors for cytokines or polypeptidic hormones.
- Non-receptor tyrosine-protein kinase TYK2.
- Protein-tyrosine phosphatases PTPN3 and PTPN4, enzymes that appear to act at junctions between the membrane and the cytoskeleton.
- Protein-tyrosine phosphatases PTPN14 and PTP-D1, PTP-RL10 and PTP2E.
- *Caenorhabditis elegans* protein phosphatase ptp-1.

Protein Subunit

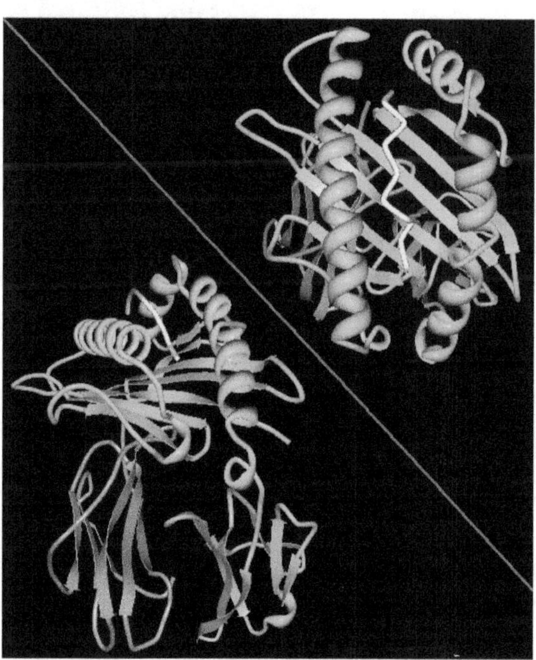

Rendering of HLA-A11 showing the α (A*1101 gene product) and β (Beta-2 microglobin) subunits. This receptor has a bound peptide (in the binding pocket) of heterologous origin that also contributes to function.

In structural biology, a protein subunit is a single protein molecule that assembles (or "*coassembles*") with other protein molecules to form a protein complex. Some naturally occurring proteins have a relatively small number of subunits and therefore described as *oligomeric*, for example hemoglobin or DNA polymerase. Others may consist from a very large number of subunits and therefore described as *multimeric*, for example microtubules and other cytoskeleton proteins. The subunits of a multimeric protein may be identical, homologous or totally dissimilar and dedicated to disparate tasks.

In some protein assemblies, one subunit may be a "catalytic subunit" that enzymatically catalyzes a reaction, whereas a "regulatory subunit" will facilitate or inhibit the activity. Although telomerase has telomerase reverse transcriptase as a catalytic subunit, regulation is accomplished by factors outside of the protein. An enzyme composed of both regulatory and catalytic subunits when assembled is often referred to as a holoenzyme. For example, class I phosphoinositide 3-kinase is composed of a p110 catalytic subunit and a p85 regulatory subunit. One subunit is made of one polypeptide chain. A polypeptide chain has one gene coding for it – meaning that a protein must have one gene for each unique subunit.

A subunit is often named with a Greek or Roman letter, and the numbers of this type of subunit in a protein is indicated by a subscript. For example, ATP synthase has a type of subunit called α. Three of these are present in the ATP synthase molecule, and is therefore designated α_3. Larger groups of subunits can also the specified, like $\alpha_3\beta_3$-hexamer and c-ring.

Subunit Vaccines

A subunit vaccine presents an antigen to the immune system without introducing viral particles, whole or otherwise. One method of production involves isolation of a specific protein from a virus and administering this by itself. A weakness of this technique is that isolated proteins can be denatured and will then become associated with antibodies different from the desired antibodies. A second method of making a subunit vaccine involves putting an antigen's gene from the targeted virus or bacterium into another virus (virus vector), yeast (yeast vector in the case of the hepatitis B vaccine or attenuated bacterium (bacterial vector) to make a recombinant virus or bacteria to serve as the important component of a recombinant vaccine (called a recombinant subunit vaccine). The recombinant vector that is genomically modified will express the antigen. The antigen (one or more subunits of protein) is extracted from the vector. Just like the highly successful subunit vaccines, the recombinant-vector-produced antigen will be of little to no risk to the patient. This is the type of vaccine currently in use for hepatitis B, and it is experimentally popular, being used to try to develop new vaccines for difficult-to-vaccinate-against viruses such as ebolavirus and HIV.

Vi capsular polysaccharide vaccine (ViCPS) is another subunit vaccine (contains the signature polysaccharide linked to the Vi capsular antigen), in this case, against typhoid caused by the Typhi serotype of Salmonella. It is also called a conjugate vaccine, in which a polysaccharide antigen has been covalently attached to a carrier protein for T-cell-dependent antigen processing (utilizing MHC II).

RNA-binding Protein

RNA-binding proteins (often abbreviated as RBPs) are proteins that bind to the double or single stranded RNA in cells and participate in forming ribonucleoprotein complexes. RBPs contain various structural motifs, such as RNA recognition motif (RRM), dsRNA binding domain, zinc finger and others. They are cytoplasmic and nuclear proteins. However, since most mature RNA is exported from the nucleus relatively quickly, most RBPs in the nucleus exist as complexes of protein and pre-mRNA called heterogeneous ribonucleoprotein particles (hnRNPs). RBPs have crucial roles in various cellular processes such as: cellular function, transport and localization. They especially play a major role in post-transcriptional control of RNAs, such as: splicing, polyadenylation, mRNA stabilization, mRNA localization and translation. Eukaryotic cells encode diverse RBPs, approximately 500 genes, with unique RNA-binding activity and protein-protein interaction. During evolution, the diversity of RBPs greatly increased with the increase in the number of introns. Diversity enabled eukaryotic cells to utilize RNA exons in various arrangements, giving rise to a unique RNP (ribonucleoprotein) for each RNA. Although RBPs have a crucial role in post-transcriptional regulation in gene expression, relatively few RBPs have been studied systematically.

Structure

Many RBPs have modular structures and are composed of multiple repeats of just a few specific basic domains that often have limited sequences. These sequences are then arranged in varying combinations to fulfill the need for diversity. A specific protein's recognition of a specific RNA has evolved through the rearrangement of these few basic domains. Each basic domain recognizes RNA, but many of these proteins require multiple copies of one of the many common domains to function.

Diversity

As nuclear RNA emerges from RNA polymerase, RNA transcripts are immediately covered with RNA-binding proteins that regulate every aspect of RNA metabolism and function including RNA biogenesis, maturation, transport, cellular localization and stability. All RBPs bind RNA, however they do so with different RNA-sequence specificities and affinities, which allows the RBPs to be as diverse as their targets and functions. These targets include mRNA, which codes for proteins, as well as a number of functional non-coding RNAs. NcRNAs almost always function as ribonucleoprotein complexes and not as naked RNAs. These non-coding RNAs include microRNAs, small interfering RNAs (siRNA), as well as splicesomal small nuclear RNAs (snRNA).

Function

RNA Processing and Modification

Alternative Splicing

Alternative splicing is a mechanism by which different forms of mature mRNAs (messengers RNAs) are generated from the same gene. It is a regulatory mechanism by which variations in the incorporation of the exons into mRNA leads to the production of more than one related protein, thus expanding possible genomic outputs. RBPs function extensively in the regulation of this process. Some binding proteins such as neuronal specific RNA-binding proteins, namely NOVA1, control the alternative splicing of a subset of hnRNA by recognizing and binding to a specific sequence in the RNA (YCAY where Y indicates pyrimidine, U or C). These proteins then recruit splicesomal proteins to this target site. SR proteins are also well known for their role in alternative splicing through the recruitment of snRNPs that form the splicesome, namely U1 snRNP and U2AF snRNP. However, RBPs are also part of the splicesome itself. The splicesome is a complex of snRNA and protein subunits and acts as the mechanical agent that removes introns and ligates the flanking exons. Other than core splicesome complex, RBPs also bind to the sites of *Cis*-acting RNA elements that influence exons inclusion or exclusion during splicing. These sites are referred to as exonic splicing enhancers (ESEs), exonic splicing silencers (ESSs), intronic splicing enhancers (ISEs) and intronic splicing silencers (ISSs) and depending on their location of binding, RBPs work as splicing silencers or enhancers.

RNA Editing

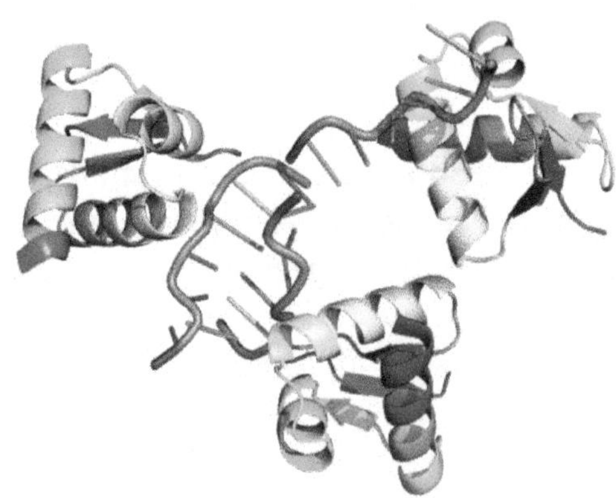

ADAR : an RNA binding protein involved in RNA editing events.

The most extensively studied form of RNA editing involves the ADAR protein. This protein functions through post-transcriptional modification of mRNA transcripts by changing the nucleotide content of the RNA. This is done through the conversion of adenosine to inosine in an enzymatic reaction catalyzed by ADAR. This process effectively changes the RNA sequence from that encoded by the genome and extends the diversity of the gene products. The majority of RNA editing occurs on non-coding regions of RNA; however, some protein-encoding RNA transcripts have been shown to be subject to editing resulting in a difference in their protein's amino acid sequence. An example of this is the glutamate receptor mRNA where glutamine is converted to arginine leading to a change in the functionality of the protein.

Polyadenylation

Polyadenylation is the addition of a "tail" of adenylate residues to an RNA transcript about 20 bases downstream of the AAUAAA sequence within the three prime untranslated region. Polyadenylation of mRNA has a strong effect on its nuclear transport, translation efficiency, and stability. All of these as well as the process of polyadenylation depend on binding of specific RBPs. All eukaryotic mRNAs with few exceptions are processed to receive 3' poly (A) tails of about 200 nucleotides. One of the necessary protein complexes in this process is CPSF. CPSF binds to the 3' tail (AAUAAA) sequence and together with another protein called poly(A)-binding protein, recruits and stimulates the activity of poly(A) polymerase. Poly(A) polymerase is inactive on its own and requires the binding of these other proteins to function properly.

Export

After processing is complete, mRNA needs to be transported from the cell nucleus to cytoplasm. This is a three-step process involving the generation of a cargo-carrier complex in the nucleus followed by translocation of the complex through the nuclear pore complex and finally release of the cargo into cytoplasm. The carrier is then subsequently recycled. TAP/NXF1:p15 heterodimer is thought to be the key player in mRNA export. Over-expression of TAP in *Xenopus laevis* frogs increases the export of transcripts that are otherwise inefficiently exported. However TAP needs adaptor proteins because it is unable interact directly with mRNA. Aly/REF protein interacts and bind to the mRNA recruiting TAP.

mRNA Localization

mRNA localization is critical for regulation of gene expression by allowing spatially regulated protein production. Through mRNA localization proteins are transcribed in their intended target site of the cell. This is especially important during early development when rapid cell cleavages give different cells various combinations of mRNA which can then lead to drastically different cell fates. RBPs are critical in the localization of this mRNA that insures proteins are only transcribed in their intended regions. One of these proteins is ZBP1. ZBP1 binds to beta-actin mRNA at the site of transcription and moves with mRNA into the cytoplasm. It then localizes this mRNA to the lamella region of several asymmetric cell types where it can then be translated. FMRP is another RBP involved in RNA localization. It was shown that in addition to other functions for FMRP in RNA metabolism, FMRP is involved in the stimulus-induced localization of several dendritic mRNAs in neuronal dendrites.

Translation

Translational regulation provides a rapid mechanism to control gene expression. Rather than controlling gene expression at the transcriptional level, mRNA is already transcribed but the recruitment of ribosomes is controlled. This allows rapid generation of proteins when a signal activates translation. ZBP1 in addition to its role in the localization of B-actin mRNA is also involved in the translational repression of beta-actin mRNA by blocking translation initiation. ZBP1 must be removed from the mRNA to allow the ribosome to properly bind and translation to begin.

RNA-binding Activity and Recognition of the RNA Sequence

RNA-binding proteins exhibit highly specific recognition of their RNA targets by recognizing their sequences and structures. Specific binding of the RNA-binding proteins allow them to distinguish their targets and regulate a variety of cellular functions via control of the generation, maturation, and lifespan of the RNA transcript. This interaction begins during transcription as some RBPs remain bound to RNA until degradation whereas others only transiently bind to RNA to regulate RNA splicing, processing, transport, and localization. In this section, three classes of the most widely studied RNA-binding domains (RNA-recognition motif, double-stranded RNA-binding motif, zinc-finger motif) will be discussed.

RNA-recognition Motif (RRM)

The RNA recognition motif, which is the most common RNA-binding motif, is a small protein domain of 75-85 amino acids that forms a four-stranded β-sheet against the two α-helices. This recognition motif exerts its role in numerous cellular functions, especially in mRNA/rRNA processing, splicing, translation regulation, RNA export, and RNA stability. Ten structures of an RRM have been identified through NMR spectroscopy and X-ray crystallography. These structures illustrate the intricacy of protein-RNA recognition of RRM as it entails RNA-RNA and protein-protein interactions in addition to protein-RNA interactions. Despite their complexity, all ten structures have some common features. All RRMs' main protein surfaces' four-stranded β-sheet was found to interact with the RNA, which usually contacts two or three nucleotides in a specific manner. In addition, strong RNA binding affinity and specificity towards variation are achieved through an interaction between the inter-domain linker and the RNA and between RRMs themselves. This plasticity of the RRM explains why RRM is the most abundant domain and why it plays an important role in various biological functions.

Double-stranded RNA-binding Motif (dsRBM)

The dsRBM, 70-75 amino-acid domain, plays a critical role in RNA processing, RNA localization, RNA interference, RNA editing, and translational repression. Although only three structures of dsRBMs have been currently discovered, all three structures possess uniting features that explain how dsRBMs only bind to dsRNA instead of dsDNA. The dsRBMs were found to interact along the RNA duplex via both α-helices and β1-β2 loop. Moreover, all three dsRBM structures make contact with the sugar-phosphate backbone of the major groove and of one minor groove, which is mediated by the β1-β2 loop along with the N-terminus region of the alpha helix 2. This interaction is a unique adaptation for the shape of an RNA double helix as it involves 2'-hydroxyls and phosphate oxygen. Despite the common structural features among dsRBMs, they exhibit distinct chemical frameworks, which permits specificity for a variety for RNA structures including stem-loops, internal loops, bulges or helices containing mismatches.

Zinc Fingers

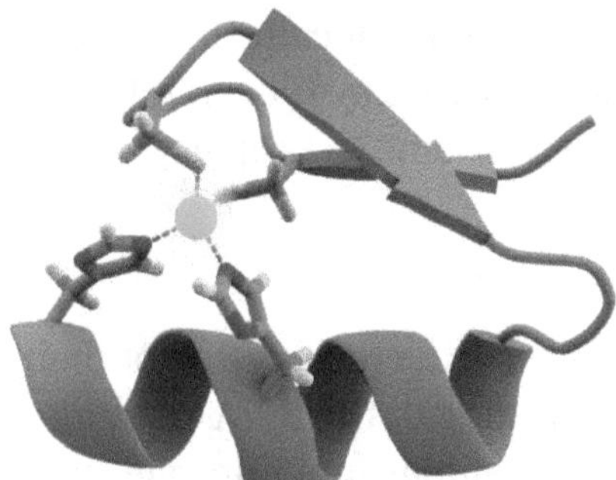

"Zinc finger" : Cartoon representation of the zinc-finger motif of proteins. The zinc ion (green) is coordinated by two histidine and two cysteine amino acid residues.

CCHH-type zinc-finger domains are the most common DNA-binding domain within the eukaryotic genome. In order to attain high sequence-specific recognition of DNA, several zinc fingers are utilized in a modular fashion. Zinc fingers exhibit ββα protein fold in which a β-hairpin and a α-helix are joined together via a Zn2+ion. Furthermore, the interaction between protein side-chains of the α-helix with the DNA bases in the major groove allows for the DNA-sequence-specific recognition. Despite its wide recognition of DNA, there has been recent discoveries that zinc fingers also have the ability to recognize RNA. In addition to CCHH zinc fingers, CCCH zinc fingers were recently discovered to employ sequence-specific recognition of single-stranded RNA through an interaction between intermolecular hydrogen bonds and Watson-Crick edges of the RNA bases. CCHH-type zinc fingers employ two methods of RNA binding. First, the zinc fingers exert non-specific interaction with the backbone of a double helix whereas the second mode allows zinc fingers to specifically recognize the individual bases that bulge out. Differing from the CCHH-type, the CCCH-type zinc finger displays another mode of RNA binding, in which single-stranded RNA is identified in a sequence-specific manner. Overall, zinc fingers can directly recognize DNA via binding to dsDNA sequence and RNA via binding to ssRNA sequence.

Role in Embryonic Development

Crawling C. elegans hermaphrodite worm

RNA-binding proteins' transcriptional and post-transcriptional regulation of RNA has a role in regulating the patterns of gene expression during development. Extensive research on the nematode *C. elegans* has identified RNA-binding proteins as essential factors during germline and early embryonic development. Their specific function involves the development of somatic tissues (neurons, hypodermis, muscles and excretory cells) as well as providing timing cues for the developmental events. Nevertheless, it is exceptionally challenging to discover the mechanism behind RBPs' function in development due to the difficulty in identifying their RNA targets. This is because most RBPs usually have multiple RNA targets. However, it is indisputable that RBPs exert a critical control in regulating developmental pathways in a concerted manner.

RBPs in Germline Development

In *Drosophila melanogaster*, Elav, Sxl and tra-2 are RNA-binding protein encoding genes that are critical in the early sex determination and the maintenance of the somatic sexual state. These genes impose effects on the post-transcriptional level by regulating sex-specific splicing in *Drosophila.* Sx1 exerts positive regulation of the feminizing gene *tra* to produce a functional tra mRNA in females. In *C. elegans*, RNA-binding proteins including FOG-1, MOG-1/-4/-5 and RNP-4 regulate germline and somatic sex determination. Furthermore, several RBPs such as GLD-1, GLD-3, DAZ-1, PGL-1 and OMA-1/-2 exert their regulatory functions during meiotic prophase progression, gametogenesis, and oocyte maturation.

RBPs in Somatic Development

In addition to RBPs' functions in germline development, post-transcriptional control also plays a significant role in somatic development. Differing from RBPs that are involved in germline and early embryo development, RBPs functioning in somatic development regulate tissue-specific alternative splicing of the mRNA targets. For instance, MEC-8 and UNC-75 containing RRM domains localize to regions of hypodermis and nervous system, respectively. Furthermore, another RRM-containing RBP, EXC-7, is revealed to localize in embryonic excretory canal cells and throughout the nervous system during somatic development.

RBPs in Neuronal Development

ZBP1 was shown to regulate dendritogenesis (dendrite formation) in hippocampal neurons. Other RNA-binding proteins involved in dendrite formation are Pumilio and Nanos, FMRP, CPEB and Staufen 1

Role of RBPs in Cancer

RBPs are emerging to play a crucial role in tumor development. In cancer, several alterations have been found in genes encoding for RNA-binding proteins. Many RBPs are differentially expressed in different cancer types for example KHDRBS1(Sam68), ELAVL1(HuR), FXR1. For some RBPs, the change in expression are related with Copy Number Variations (CNV), for example CNV gains of ESRP1, CELF3 in breast cancer, RBM24 in liver cancer, IGF2BP2, IGF2BP3 in lung cancer or CNV losses of KHDRBS2 in lung cancer. Some expression changes are cause due to protein affecting mutations on these RBPs for example SF3B1, SRSF2, RBM10, U2AF1, SF3B1, PPRC1, RBMXL1, HNRNPCL1 etc. Several studies have related this change in expression of RBPs to aberrant alternative splicing in cancer.

Current Research

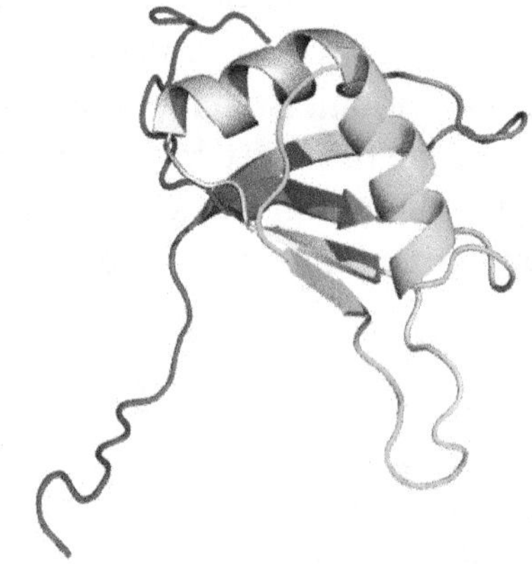

"CIRBP" : Structure of the CIRBP protein.

As RNA-binding proteins exert significant control over numerous cellular functions, they have been a popular area of investigation for many researchers. Due to its importance in the biological field, numerous discoveries regarding RNA-binding proteins' potentials have been recently unveiled. Recent development in experimental identification of RNA-binding proteins has extended the number of RNA-binding proteins significantly

RNA-binding protein Sam68 controls the spatial and temporal compartmentalization of RNA metabolism to attain proper synaptic function in dendrites. Loss of Sam68 results in abnormal posttranscriptional regulation and ultimately leads to neurological disorders such as fragile X-associated tremor/ataxia syndrome. Sam68 was found to interact with the mRNA encoding β-actin, which regulates the synaptic formation of the dendritic spines with its cytoskeletal components. Therefore, Sam68 plays a critical role in regulating synapse number via control of postsynaptic β-actin mRNA metabolism.

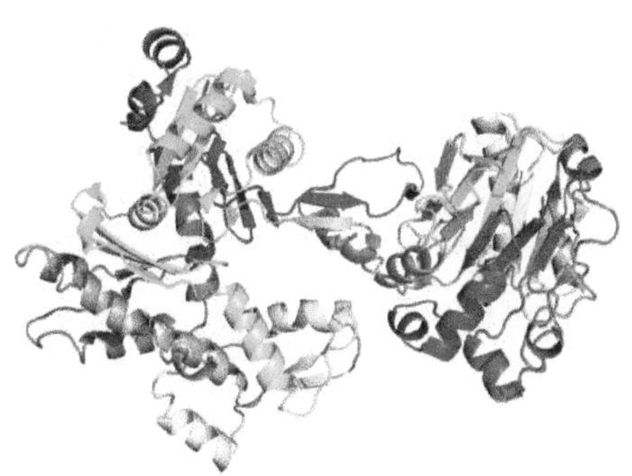

"Beta-actin" : Structure of the ACTB protein.

Neuron-specific CELF family RNA-binding protein UNC-75 specifically binds to the UUGUUGUGUUGU mRNA stretch via its three RNA recognition motifs for the exon 7a selection in *C. elegans*' neuronal cells. As exon 7a is skipped due to its weak splice sites in non-neuronal cells, UNC-75 was found to specifically activate splicing between exon 7a and exon 8 only in the neuronal cells.

The cold inducible RNA binding protein CIRBP plays a role in controlling the cellular response upon confronting a variety of cellular stresses, including short wavelength ultraviolet light, hypoxia, and hypothermia. This research yielded potential implications for the association of disease states with inflammation.

Serine-arginine family of RNA-binding protein Slr1 was found exert control on the polarized growth in Candida albicans. Slr1 mutations in mice results in decreased filamentation and reduces damage to epithelial and endothelial cells that leads to extended survival rate compared to the Slr1 wild-type strains. Therefore, this research reveals that SR-like protein Slr1 plays a role in instigating the hyphal formation and virulence in *C. albicans*.

Pleckstrin Homology Domain

Pleckstrin homology domain (PH domain) is a protein domain of approximately 120 amino acids that occurs in a wide range of proteins involved in intracellular signaling or as constituents of the cytoskeleton.

This domain can bind phosphatidylinositol lipids within biological membranes (such as phosphatidylinositol (3,4,5)-trisphosphate and phosphatidylinositol (4,5)-bisphosphate), and proteins such as the βγ-subunits of heterotrimeric G proteins, and protein kinase C. Through these interactions, PH domains play a role in recruiting proteins to different membranes, thus targeting them to appropriate cellular compartments or enabling them to interact with other components of the signal transduction pathways.

Lipid Binding Specificity

Individual PH domains possess specificities for phosphoinositides phosphorylated at different sites within the inositol ring, e.g., some bind phosphatidylinositol (4,5)-bisphosphate but not phosphatidylinositol (3,4,5)-trisphosphate or phosphatidylinositol (3,4)-bisphosphate, while others may possess the requisite affinity. This is important because it makes the recruitment of different PH domain containing proteins sensitive to the activities of enzymes that either phosphorylate or dephosphorylate these sites on the inositol ring, such as phosphoinositide 3-kinase or PTEN, respectively. Thus, such enzymes exert a part of their effect on cell function by modulating the localization of downstream signaling proteins that possess PH domains that are capable of binding their phospholipid products.

Structure

The 3D structure of several PH domains has been determined. All known cases have a common structure consisting of two perpendicular anti-parallel beta sheets, followed by a C-terminal amphipathic helix. The loops connecting the beta-strands differ greatly in length, making the PH domain relatively difficult to detect while providing the source of the domain's specificity. The only conserved residue among PH domains is a single tryptophan located within the alpha helix that serves to nucleate the core of the domain.

Proteins Containing PH Domain

PH domains can be found in many different proteins, such as OSBP or ARF. Recruitment to the Golgi in this case is dependent on both PtdIns and ARF. A large number of PH domains have poor affinity for phosphoinositides and are hypothesized to function as protein binding domains. A Genome-wide look in *Saccharomyces cerevisiae* showed that most of the 33 yeast PH domains are indeed promiscuous in binding to phosphoinositides, while only one (Num1-PH) behaved highly specific . Proteins reported to contain PH domains belong to the following families:

- Pleckstrin, the protein where this domain was first detected, is the major substrate of protein kinase C in platelets. Pleckstrin is one of the rare proteins to contain two PH domains.
- Ser/Thr protein kinases such as the Akt/Rac family, the beta-adrenergic receptor kinases, the mu isoform of PKC and the trypanosomal NrkA family.
- Tyrosine protein kinases belonging to the Btk/Itk/Tec subfamily.
- Insulin receptor substrate 1 (IRS-1).
- Regulators of small G-proteins like guanine nucleotide releasing factor GNRP (Ras-GRF) (which contains 2 PH domains), guanine nucleotide exchange proteins like vav, dbl, SoS and *S. cerevisiae* CDC24, GTPase activating proteins like rasGAP and BEM2/IPL2, and the human break point cluster protein bcr.
- Cytoskeletal proteins such as dynamin, *Caenorhabditis elegans* kinesin-like protein unc-104, spectrin beta-chain, syntrophin (2 PH domains), and S. cerevisiae nuclear migration protein NUM1.
- Mammalian phosphatidylinositol-specific phospholipase C (PI-PLC) isoforms gamma and delta. Isoform gamma contains two PH domains, the second one is split into two parts separated by about 400 residues.
- Oxysterol-binding proteins OSBP, S. cerevisiae OSH1 and YHR073w.

- Mouse protein citron, a putative rho/rac effector that binds to the GTP-bound forms of rho and rac.
- Several S. cerevisiae proteins involved in cell cycle regulation and bud formation like BEM2, BEM3, BUD4 and the BEM1-binding proteins BOI2 (BEB1) and BOI1 (BOB1).
- C. elegans protein MIG-10.
- Ceramide kinase, a lipid kinase that phosphorylates ceramides to ceramide-1-phosphate.

Subfamilies

- Spectrin/pleckstrin-like InterPro: *IPR001605*

Examples

Human genes encoding proteins containing this domain include:

- ABR, ADRBK1, ADRBK2, AFAP, AFAP1, AFAP1L1, AFAP1L2, AKAP13, AKT1, AKT2, AKT3, ANLN, APBB1IP, APPL1, APPL2, ARHGAP10, ARHGAP12, ARHGAP15, ARHGAP21, ARHGAP22, ARHGAP23, ARHGAP24, ARHGAP25, ARHGAP26, ARHGAP27, ARHGAP9, ARHGEF16, ARHGEF18, ARHGEF19, ARHGEF2, ARHGEF3, ARHGEF4, ARHGEF5, ARHGEF6, ARHGEF7, ARHGEF9, ASEF2,
- BMX, BTK,
- C20orf42, C9orf100, CADPS, CADPS2, CDC42BPA, CDC42BPB, CDC42BPG, CENTA1, CENTA2, CENTB1, CENTB2, CENTB5, CENTD1, CENTD2, CENTD3, CENTG1, CENTG2, CENTG3, CIT, CNKSR1, CNKSR2, COL4A3BP, CTGLF1, CTGLF2, CTGLF3, * CTGLF4, CTGLF5, CTGLF6,
- DAB2IP, DAPP1, DDEF1, DDEF2, DDEFL1, DEF6, DEPDC2, DGKD, DGKH, DGKK, DNM1, DNM2, DNM3, DOCK10, DOCK11, DOCK9, DOK1, DOK2, DOK3, DOK4, DOK5, DOK6, DTGCU2,
- EXOC8,
- FAM109A, FAM109B, FARP1, FARP2, FGD1, FGD2, FGD3, FGD4, FGD5, FGD6,
- GAB1, GAB2, GAB3, GAB4, GRB10, GRB14, GRB7,
- IRS1, IRS2, IRS4, ITK, ITSN1, ITSN2,
- KALRN, KIF1A, KIF1B, KIF1Bbeta,
- MCF2, MCF2L, MCF2L2, MRIP, MYO10,
- NET1, NGEF,
- OBPH1, OBSCN, OPHN1, OSBP, OSBP2, OSBPL10, OSBPL11, OSBPL3, OSBPL5, OSBPL6, OSBPL7, OSBPL8, OSBPL9,
- PHLDA2, PHLDA3, PHLDB1, PHLDB2, PHLPP, PIP3-E, PLCD1, PLCD4, PLCG1, PLCG2, PLCH1, PLCH2, PLCL1, PLCL2, PLD1, PLD2, PLEK, PLEK2, PLEKHA1, PLEKHA2,

PLEKHA3, PLEKHA4, PLEKHA5, PLEKHA6, PLEKHA7, PLEKHA8, PLEKHB1, PLEKHB2, PLEKHC1, PLEKHF1, PLEKHF2, PLEKHG1, PLEKHG2, PLEKHG3, PLEKHG4, PLEKHG5, PLEKHG6, PLEKHH1, PLEKHH2, PLEKHH3, PLEKHJ1, PLEKHK1, PLEKHM1, PLEKHM2, PLEKHO1, PLEKHQ1, PREX1, PRKCN, PRKD1, PRKD2, PRKD3, PSCD1, PSCD2, PSCD3, PSCD4, PSD, PSD2, PSD3, PSD4, RALGPS1, RALGPS2, RAPH1,

- RASA1, RASA2, RASA3, RASA4, RASAL1, RASGRF1, RGNEF, ROCK1, ROCK2, RTKN,
- SBF1, SBF2, SCAP2, SGEF, SH2B, SH2B1, SH2B2, SH2B3, SH3BP2, SKAP1, SKAP2, SNTA1, SNTB1, SNTB2, SOS1, SOS2, SPATA13, SPNB4, SPTBN1, SPTBN2, SPTBN4, SPTBN5, STAP1, SWAP70, SYNGAP1,
- TBC1D2, TEC, TIAM1, TRIO, TRIOBP, TYL,
- URP1, URP2,
- VAV1, VAV2, VAV3, VEPH1

References

- Appasani, Krishnarao (2008). MicroRNAs: From Basic Science to Disease Biology. Cambridge University Press. p. 485. ISBN 978-0-521-86598-2. Retrieved May 12, 2013.

5

Methods and Techniques in Protein-Protein Interaction

The method of separating biochemical mixtures is known as affinity chromatography whereas the chip-sequencing is the method used to analyze the interactions between proteins and DNA. Some of the other techniques discussed in the content are protein footprinting, microscale thermophoresis, two-hybrid screening, phage display and chromatin immuneprecipition.

Affinity Chromatography

Affinity chromatography is a method of separating biochemical mixtures based on a highly specific interaction such as that between antigen and antibody, enzyme and substrate, or receptor and ligand.

Uses

Affinity chromatography can be used to:

- Purify and concentrate a substance from a mixture into a buffering solution
- Reduce the amount of a substance in a mixture
- Discern what biological compounds bind to a particular substance
- Purify and concentrate an enzyme solution.

Principle

The stationary phase is typically a gel matrix, often of agarose; a linear sugar molecule derived from algae. Usually the starting point is an undefined heterogeneous group of molecules in solution, such as a cell lysate, growth medium or blood serum. The molecule of interest will have a well known and defined property, and can be exploited during the affinity purification process. The process itself can be thought of as an entrapment, with the target molecule becoming trapped on a solid or stationary phase or medium. The other molecules in the mobile phase will not become trapped as they do not possess this property. The stationary phase can then be removed from the mixture, washed and the target molecule released from the entrapment in a process known as elution. Possibly the most common use of affinity chromatography is for the purification of recombinant proteins.

Batch and Column Setups

Binding to the solid phase may be achieved by column chromatography whereby the solid medium is packed onto a column, the initial mixture run through the column to allow setting, a wash buffer run through the column and the elution buffer subsequently applied to the column and collected. These steps are usually done at ambient pressure. Alternatively, binding may be achieved using a batch treatment, for example, by adding the initial mixture to the solid phase in a vessel, mixing, separating the solid phase, removing the liquid phase, washing, re-centrifuging, adding the elution buffer, re-centrifuging and removing the eluate.

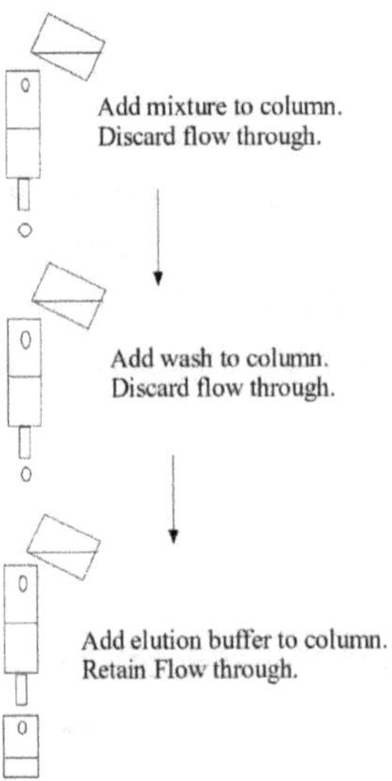

Column chromatography

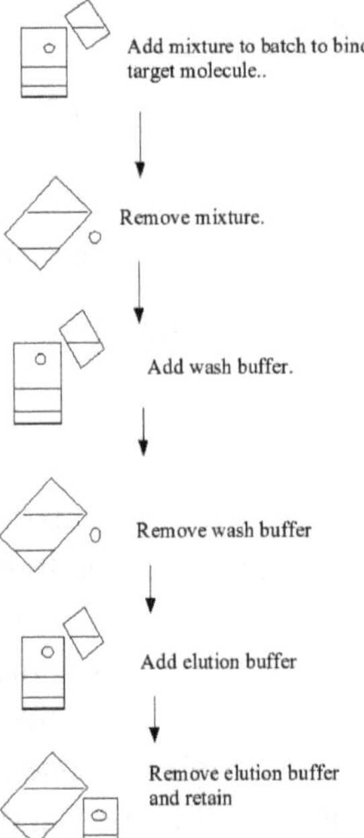

Batch chromatography

Sometimes a hybrid method is employed such that the binding is done by the batch method, but the solid phase with the target molecule bound is packed onto a column and washing and elution are done on the column.

A third method, expanded bed adsorption, which combines the advantages of the two methods mentioned above, has also been developed. The solid phase particles are placed in a column where liquid phase is pumped in from the bottom and exits at the top. The gravity of the particles ensure that the solid phase does not exit the column with the liquid phase.

Affinity columns can be eluted by changing salt concentrations, pH, pI, charge and ionic strength directly or through a gradient to resolve the particles of interest.

More recently, setups employing more than one column in series have been developed. The advantage compared to single column setups is that the resin material can be fully loaded, since non-binding product is directly passed on to a consecutive column with fresh column material. The resin costs per amount of produced product can thus be drastically be reduced. Since one column can always be eluted and regenerated while the other column is loaded, already two columns are sufficient to make full use of the advantages. Additional columns can give additional flexibility for elution and regeneration times, at the cost of additional equipment and resin costs.

Specific Uses

Affinity chromatography can be used in a number of applications, including nucleic acid purification, protein purification from cell free extracts, and purification from blood.

Various Affinity Media

Many different affinity media exist for a variety of possible uses. Briefly, they are (generalized):

- Activated/Functionalized – Works as a functional spacer, support matrix, and eliminates handling of toxic reagents.
- Amino Acid – Used with a variety of serum proteins, proteins, peptides, and enzymes, as well as rRNA and dsDNA.
- Avidin Biotin – Used in the purification process of biotin/avidin and their derivatives.
- Carbohydrate Bonding – Most often used with glycoproteins or any other carbohydrate-containing substance.
- Carbohydrate – Used with lectins, glycoproteins, or any other carbohydrate metabolite protein.
- Dye Ligand – This media is nonspecific, but mimics biological substrates and proteins.
- Glutathione – Useful for separation of GST tagged recombinant proteins.
- Heparin – This media is a generalized affinity ligand, and it is most useful for separation of plasma coagulation proteins, along with nucleic acid enzymes and lipases.
- Hydrophobic Interaction – Most commonly used to target free carboxyl groups and proteins.
- Immunoaffinity – Detailed below, this method utilizes antigens' and antibodies' high specificity to separate.

- Immobilized Metal Affinity Chromatography – Detailed further below, this method uses interactions between metal ions and proteins (usually specially tagged) to separate.
- Nucleotide/Coenzyme – Works to separate dehydrogenases, kinases, and transaminases.
- Nucleic Acid – Functions to trap mRNA, DNA, rRNA, and other nucleic acids/oligonucleotides.
- Protein A/G – This method is used to purify immunoglobulins.
- Speciality – Designed for a specific class or type of protein/coenzyme, this type of media will only work to separate a specific protein or coenzyme.

Immunoaffinity

Another use for the procedure is the affinity purification of antibodies from blood serum. If serum is known to contain antibodies against a specific antigen (for example if the serum comes from an organism immunized against the antigen concerned) then it can be used for the affinity purification of that antigen. This is also known as Immunoaffinity Chromatography. For example if an organism is immunised against a GST-fusion protein it will produce antibodies against the fusion-protein, and possibly antibodies against the GST tag as well. The protein can then be covalently coupled to a solid support such as agarose and used as an affinity ligand in purifications of antibody from immune serum.

For thoroughness the GST protein and the GST-fusion protein can each be coupled separately. The serum is initially allowed to bind to the GST affinity matrix. This will remove antibodies against the GST part of the fusion protein. The serum is then separated from the solid support and allowed to bind to the GST-fusion protein matrix. This allows any antibodies that recognize the antigen to be captured on the solid support. Elution of the antibodies of interest is most often achieved using a low pH buffer such as glycine pH 2.8. The eluate is collected into a neutral tris or phosphate buffer, to neutralize the low pH elution buffer and halt any degradation of the antibody's activity. This is a nice example as affinity purification is used to purify the initial GST-fusion protein, to remove the undesirable anti-GST antibodies from the serum and to purify the target antibody.

A simplified strategy is often employed to purify antibodies generated against peptide antigens. When the peptide antigens are produced synthetically, a terminal cysteine residue is added at either the N- or C-terminus of the peptide. This cysteine residue contains a sulfhydryl functional group which allows the peptide to be easily conjugated to a carrier protein (e.g. Keyhole limpet hemocyanin (KLH)). The same cysteine-containing peptide is also immobilized onto an agarose resin through the cysteine residue and is then used to purify the antibody.

Most monoclonal antibodies have been purified using affinity chromatography based on immunoglobulin-specific Protein A or Protein G, derived from bacteria.

Immobilized Metal ion Affinity Chromatography

Immobilized metal ion affinity chromatography (IMAC) is based on the specific coordinate covalent bond of amino acids, particularly histidine, to metals. This technique works by allowing proteins with an affinity for metal ions to be retained in a column containing immobilized metal ions, such as cobalt, nickel, copper for the purification of histidine containing proteins or peptides, iron, zinc or gallium for the purification of phosphorylated proteins or peptides. Many naturally occurring proteins do not have an affinity for metal ions, therefore recombinant DNA technology can be used to introduce such a protein tag into the relevant gene. Methods used to elute the protein of interest include changing the pH, or adding a competitive molecule, such as imidazole.

A chromatography column containing nickel-agarose beads used for purification of proteins with histidine tags

Recombinant Proteins

Possibly the most common use of affinity chromatography is for the purification of recombinant proteins. Proteins with a known affinity are protein tagged in order to aid their purification. The protein may have been genetically modified so as to allow it to be selected for affinity binding; this is known as a fusion protein. Tags include glutathione-S-transferase (GST), hexahistidine (His), and maltose binding protein (MBP). Histidine tags have an affinity for nickel or cobalt ions which have been immobilized by forming coordinate covalent bonds with a chelator incorporated in the stationary phase. For elution, an excess amount of a compound able to act as a metal ion ligand, such as imidazole, is used. GST has an affinity for glutathione which is commercially available immobilized as glutathione agarose. During elution, excess glutathione is used to displace the tagged protein.

Lectins

Lectin affinity chromatography is a form of affinity chromatography where lectins are used to separate components within the sample. Lectins, such as Concanavalin A are proteins which can bind specific carbohydrate (sugar) molecules. The most common application is to separate glycoproteins from non-glycosylated proteins, or one glycoform from another glycoform.

ChIP-sequencing

ChIP-sequencing, also known as ChIP-seq, is a method used to analyze protein interactions with DNA. ChIP-seq combines chromatin immunoprecipitation (ChIP) with massively parallel DNA sequencing to identify the binding sites of DNA-associated proteins. It can be used to map global binding sites precisely for any protein of interest. Previously, ChIP-on-chip was the most common technique utilized to study these protein–DNA relations.

Uses

ChIP-seq is used primarily to determine how transcription factors and other chromatin-associated proteins influence phenotype-affecting mechanisms. Determining how proteins interact with DNA to regulate gene expression is essential for fully understanding many biological processes and disease states. This epigenetic information is complementary to genotype and expression analysis. ChIP-seq technology is currently seen primarily as an alternative to ChIP-chip which requires a hybridization array. This necessarily introduces some bias, as an array is restricted to a fixed number of probes. Sequencing, by contrast, is thought to have less bias, although the sequencing bias of different sequencing technologies is not yet fully understood.

Specific DNA sites in direct physical interaction with transcription factors and other proteins can be isolated by chromatin immunoprecipitation. ChIP produces a library of target DNA sites bound to a protein of interest *in vivo*. Massively parallel sequence analyses are used in conjunction with whole-genome sequence databases to analyze the interaction pattern of any protein with DNA, or the pattern of any epigenetic chromatin modifications. This can be applied to the set of ChIP-able proteins and modifications, such as transcription factors, polymerases and transcriptional machinery, structural proteins, protein modifications, and DNA modifications. As an alternative to the dependence on specific antibodies, different methods have been developed to find the superset of all nucleosome-depleted or nucleosome-disrupted active regulatory regions in the genome, like DNase-Seq and FAIRE-Seq.

Workflow of ChIP-sequencing

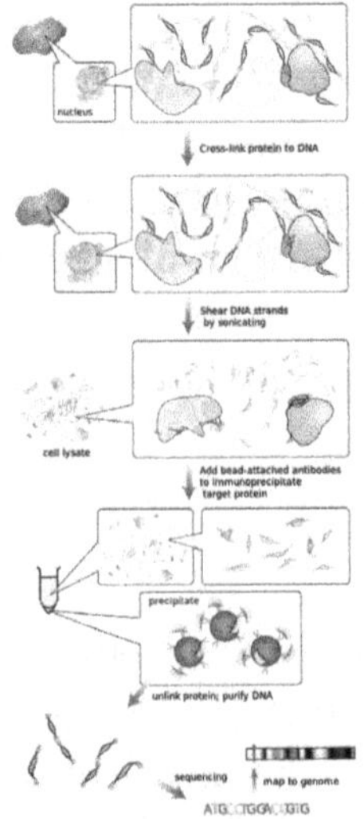

ChIP-sequencing workflow

ChIP

ChIP is a powerful method to selectively enrich for DNA sequences bound by a particular protein in living cells. However, the widespread use of this method has been limited by the lack of a sufficiently robust method to identify all of the enriched DNA sequences. The ChIP process enriches specific crosslinked DNA-protein complexes using an antibody against the protein of interest. Oligonucleotide adaptors are then added to the small stretches of DNA that were bound to the protein of interest to enable mas-sively parallel sequencing.

Sequencing

After size selection, all the resulting ChIP-DNA fragments are sequenced simultaneously using a genome sequencer. A single sequencing run can scan for genome-wide associations with high resolution, meaning that features can be located precisely on the chromosomes. ChIP-chip, by contrast, requires large sets of tiling arrays for lower resolution.

There are many new sequencing methods used in this sequencing step. Some technologies that analyze the sequences can use cluster amplification of adapter-ligated ChIP DNA fragments on a solid flow cell substrate to create clusters of approximately 1000 clonal copies each. The resulting high density array of template clusters on the flow cell surface is sequenced by a Genome analyzing program. Each template cluster undergoes sequencing-by-synthesis in parallel using novel fluorescently labelled reversible terminator nucleotides. Templates are sequenced base-by-base during each read. Then, the data collection and analysis software aligns sample sequences to a known genomic sequence to identify the ChIP-DNA fragments.

Sensitivity

Sensitivity of this technology depends on the depth of the sequencing run (i.e. the number of mapped sequence tags), the size of the genome and the distribution of the target factor. The sequencing depth is directly correlated with cost. If abundant binders in large genomes have to be mapped with high sensitivity, costs are high as an enormously high number of sequence tags will be required. This is in contrast to ChIP-chip in which the costs are not correlated with sensitivity.

Unlike microarray-based ChIP methods, the precision of the ChIP-seq assay is not limited by the spacing of predetermined probes. By integrating a large number of short reads, highly precise binding site localization is obtained. Compared to ChIP-chip, ChIP-seq data can be used to locate the binding site within few tens of base pairs of the actual protein binding site. Tag densities at the binding sites are a good indicator of protein–DNA binding affinity, which makes it easier to quantify and compare binding affinities of a protein to different DNA sites.

Current Research

STAT1 DNA association: ChIP-seq was used to study STAT1 targets in HeLA S3 cells. The performance of ChIP-seq was then compared to the alternative protein–DNA interaction methods of ChIP-PCR and ChIP-chip.

Nucleosome Architecture of Promoters: Using ChIP-seq, it was determined that Yeast genes seem to have a minimal nucleosome-free promoter region of 150bp in which RNA polymerase can initiate transcription.

Transcription factor conservation: ChIP-seq was used to compare conservation of TFs in the forebrain and heart tissue in embryonic mice. The authors identified and validated the heart functionality of transcription enhancers, and determined that transcription enhancers for the heart are less conserved than those for the forebrain during the same developmental stage.

Genome-wide ChIP-seq: ChIP-sequencing was completed on the worm C. elegans to explore genome-wide binding sites of 22 transcription factors. Up to 20% of the annotated candidate genes were assigned to transcription factors. Several transcription factors were assigned to non-coding RNA regions and may be subject to developmental or environmental variables. The functions of some of the transcription factors were also identified. Some of the transcription factors regulate genes that control other transcription factors. These genes are not regulated by other factors. Most transcription factors serve as both targets and regulators of other factors, demonstrating a network of regulation.

Inferring regulatory network: ChIP-seq signal of Histone modification were shown to be more correlated with transcription factor motifs at promoters in comparison to RNA level. Hence author proposed that using histone modification ChIP-seq would provide more reliable inference of gene-regulatory networks in comparison to other methods based on expression.

ChIP-seq offers an alternative to ChIP-chip. STAT1 experimental ChIP-seq data have a high degree of similarity to results obtained by ChIP-chip for the same type of experiment, with >64% of peaks in shared genomic regions. Because the data are sequence reads, ChIP-seq offers a rapid analysis pipeline (as long as a high-quality genome sequence is available for read mapping, and the genome doesn't have repetitive content that confuses the mapping process) as well as the potential to detect mutations in binding-site sequences, which may directly support any observed changes in protein binding and gene regulation.

Computational Analysis

As with many high-throughput sequencing approaches, ChIP-seq generates extremely large data sets, for which appropriate computational analysis methods are required. To predict DNA-binding sites from ChIP-seq read count data, peak calling methods have been developed. The most popular method is MACS which empirically models the shift size of ChIP-Seq tags, and uses it to improve the spatial resolution of predicted binding sites.

Another relevant computational problem is Differential peak calling, which identifies significant differences in two ChIP-seq signals from distinct biological conditions. Differential peak callers segment two ChIP-seq signals and identify differential peaks using Hidden Markov Models. Examples for two-stage differential peak callers are ChIPDiff and ODIN.

Protein-fragment Complementation Assay

A Protein-fragment complementation assay, or PCA, is a method for the identification of protein–protein interactions, especially in the field of proteomics. In the PCA, the proteins of interest ("bait" and "prey") are each covalently linked to incomplete fragments of a third protein (e.g. DHFR, which acts as a "reporter"). Interaction between the bait and the prey proteins brings the fragments of the reporter protein in close enough proximity to allow them to form a functional reporter protein whose activity can be measured. This principle can be applied to many different reporter proteins and is also the basis for the yeast two-hybrid system, an archetypical PCA assay.

Split Protein Assays

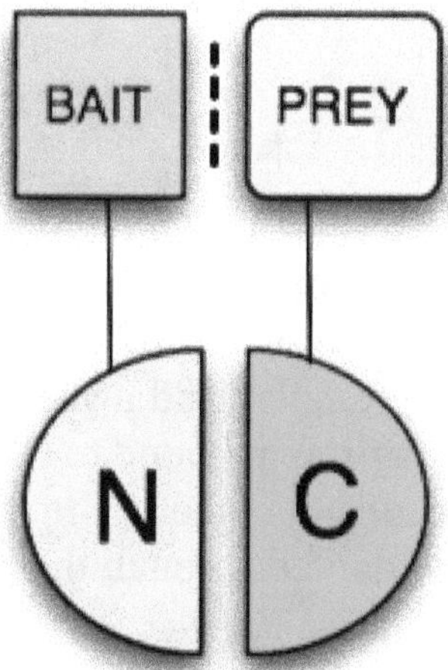

General principle of the protein complementation assay: a protein is split into two (N- and C-terminal) halves and reconstituted by two interacting proteins (here called "bait" and "prey" because a bait protein can be used to find an interacting prey protein). The activity of the reconstituted protein should be easily detectable, e.g. as in the Green Fluorescent Protein (GFP).

Any protein that can be split into two parts and reconstituted non-covalently may be used in a PCA. The two parts just have to be brought together by other interacting proteins fused to them (often called "bait" and "prey" because a bait protein can be used to find a prey protein). The protein that produces a detectable readout is called "reporter". Usually enzymes which confer resistance to antibiotics, such as Dihydrofolate reductase or Beta-lactamase, or proteins that give colorimetric or fluorescent signals are used as reporters. When fluorescent proteins are reconstituted the PCA is called Bimolecular fluorescence complementation assay. The following proteins have been used in split protein PCAs:

- Beta-lactamase
- Dihydrofolate reductase (DHFR)
- Focal adhesion kinase (FAK)
- Gal4, a yeast transcription factor (as in the classical yeast two-hybrid system)
- GFP (split-GFP), e.g. EGFP (enhanced green fluorescent protein)
- Horseradish peroxidase
- Infrared fluorescent protein IFP1.4, an engineered chromophore-binding domain (CBD) of a bacteriophytochrome from *Deinococcus radiodurans*

- LacZ (beta-galactosidase)
- Luciferase, including ReBiL (recombinase enhanced bimolecular luciferase) and the commercial products NanoLuc and NanoBIT from Promega.
- TEV (Tobacco etch virus protease)
- Ubiquitin

Protein Footprinting

Protein footprinting is a term used to refer to a method of biochemical analysis that investigates protein structure, assembly, and interactions within a larger macromolecular assembly. It was originally coined in reference to the use of limited proteolysis to investigate contact sites within a monoclonal antibody - protein antigen complex and a year later to examine the protection from hydroxyl radical cleavage conferred by a protein bound to DNA within a DNA-protein complex. In DNA footprinting the protein is envisioned to make an imprint (or footprint) at a particular point of interaction. This latter method was adapted through the direct treatment of proteins and their complexes with hydroxyl radicals.

Hydroxyl Radical Protein Footprinting

Time-resolved hydroxyl radical protein footprinting employing mass spectrometry analysis was developed in the late 1990s in synchrotron radiolysis studies. The same year, these authors reported on the use of an electrical discharge source to effect the oxidation of proteins on millisecond timescales as proteins pass from the electrosprayed solution into the mass spectrometer. These approaches have since been used to determine protein structures, protein folding, protein dynamics, and protein–protein interactions.

Unlike nucleic acids, proteins oxidize rather than cleave on these timescales. Analysis of the products by mass spectrometry reveals that proteins to are oxidized in a limited manner (some 10–30% of total protein) at a number of amino acid side chains across the proteins. The rate or level of oxidation at the reactive amino acid side chains (Met, Cys, Trp, Tyr, Phe, His, Pro and Leu) provides a measure of their accessibility to the bulk solvent. The mechanisms of side chain oxidation was explored by performing the radiolysis reactions in ^{18}O-labeled water.

Producing OH Radicals

A critical feature of these experiments is the need to expose proteins to hydroxyl radicals for limited timescales on the order of 1–50 ms inducing 10-30% oxidation of total protein. A further requirement is to generate hydroxyl radicals from the bulk solvent (i.e. water) (equations 1 and 2) not hydrogen peroxide which can remain to oxidize proteins even without other stimuli.

$$H_2O \rightarrow H_2O^{+\cdot} + e^- + H_2O^*$$

$$H_2O^{+\cdot} + H_2O \rightarrow H_3O^+ + OH^\cdot$$

Hydroxyl radicals can be produced in solution by an electrical discharge within a conventional atmospheric pressure electrospray ionization (ESI) source. When a high voltage difference (~8 keV) is held between an electrospray needle and a sampling orifice to the mass analyzer, radicals can be produced in solution at the electrospray needle tip. This method was the first employed to apply protein footprinting to the study of a protein complex.

Method

The exposure of proteins to a "white" X-ray beam of synchrotron light or an electrical discharge for tens of milliseconds provides sufficient oxidative modification to the surface amino acid side chains without damage to the protein structure. These products can be easily detected and quantified by mass spectrometry. By adjusting the time for radiolysis or which protein ions spend in the discharge source, a time-resolved approach is possible which is valuable for the study of protein dynamics.

Analysis

A computer program (PROXIMO) has also been written to help model protein complexes using data from the RP-MS/Protein footprinting approach. RP-MS/Protein footprinting studies of protein complexes can also employ computational approaches to assist with this modeling.

Applications

The application of ion mobility mass spectrometry has conclusively demonstrated that the conditions employed in RP-MS/Protein footprinting experiments do not alter the structure of proteins.

Other studies have extended the method to study early onset protein damage given the radical basis of the method and the significance of oxygen based radicals in the pathogenesis of many diseases including neurological disorders and even blindness.

Microscale Thermophoresis

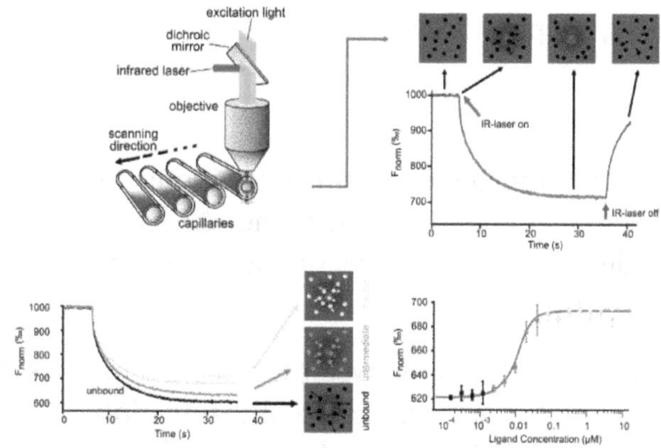

Principle of the MST Technology: MST is performed in thin capillaries in free solution thus providing close-to-native conditions (immobilization free in any buffer, even in complex bioliquids) and a maintenance free instrument. When performing a MST experiment, a microscopic temperature gradient is induced by an infrared laser, and the directed movement of molecules is detected and quantified. Thermophoresis, the movement of the molecule in the temperature gradient, depends on three parameters that typically change upon interaction. Thus, the thermophoresis signal is plotted against the ligand concentration to obtain a dose-response curve, from which the binding affinity can be deduced.

Microscale thermophoresis (MST) is a technology for the interaction analysis of biomolecules. Microscale thermophoresis is the directed movement of particles in a microscopic temperature gradient. Any change of the hydration shell of biomolecules due to changes in their structure/conformation results in a relative change of the movement along the temperature gradient and is used to determine binding affinities. MST allows measurement of interactions directly in solution without the need of immobilization to a surface (immobilization-free technology).

Applications

Affinity

- between any kind of biomolecules including proteins, DNA, RNA, peptides, small molecules, fragments and ions
- for interactions with high molecular weight complexes, large molecule assemblies, even with liposomes, vesicles, nanodiscs, nanoparticles and viruses
- in any buffer, including serum and cell lysate
- in competition experiments (for example with subtrate and inhibitors)

Stoichiometry

Thermodynamic parameters

Additional information

- Sample property (homogeneity, aggregation, stability)
- Multiple binding sites, cooperativity

Technology

MST is based on the directed movement of molecules along temperature gradients, an effect termed thermophoresis. A spatial temperature difference ΔT leads to a depletion of molecule concentration in the region of elevated temperature, quantified by the Soret coefficient S_T: $c_{hot}/c_{cold} = \exp(-S_T \Delta T)$

Thermophoresis depends on the interface between molecule and solvent. Under constant buffer conditions, thermophoresis probes the size, charge and solvation entropy of the molecules. The thermophoresis of a fluorescently labeled molecule A typically differs significantly from the thermophoresis of a molecule-target complex AT due to size, charge and solvation entropy differences. This difference in the molecule's thermophoresis is used to quantify the binding in titration experiments under constant buffer conditions.

The thermophoretic movement of the fluorescently labelled molecule is measured by monitoring the fluorescence distribution F inside a capillary . The microscopic temperature gradient is generated by an IR-Laser, which is focused into the capillary and is strongly absorbed by water. The temperature of the aqueous solution in the laser spot is raised by up to $\Delta T=5$ K. Before the IR-Laser is switched on a homogeneous fluorescence distribution F_{cold} is observed inside the capillary. When the IR-Laser is switched on, two effects, separated by their time-scales, contribute to the new fluorescence distribution F_{hot}. The thermal relaxation time is fast and induces a binding-dependent drop in the fluorescence of the dye due to its local environmental-dependent response to the temperature jump. On the slower diffusive time scale (10 s), the molecules move from the locally heated region to the outer cold regions. The local concentration of molecules decreases in the heated region until it reaches a steady-state distribution.

While the mass diffusion D dictates the kinetics of depletion, S_T determines the steady-state concentration ratio $c_{hot}/c_{cold}=\exp(-S_T\,\Delta T) \approx 1-S_T\,\Delta T$ under a temperature increase ΔT. The normalized fluorescence $F_{norm}=F_{hot}/F_{cold}$ measures mainly this concentration ratio, in addition to the temperature jump $\partial F/\partial T$. In the linear approximation we find: $F_{norm}=1+(\partial F/\partial T-S_T)\Delta T$. Due to the linearity of the fluorescence intensity and the thermophoretic depletion, the normalized fluorescence from the unbound molecule $F_{norm}(A)$ and the bound complex $F_{norm}(AT)$ superpose linearly. By denoting x the fraction of molecules bound to targets, the changing fluorescence signal during the titration of target T is given by: $F_{norm}=(1-x)\,F_{norm}(A)+x\,F_{norm}(AT)$.

Quantitative binding parameters are obtained by using a serial dilution of the binding substrate. By plotting F_{norm} against the logarithm of the different concentrations of the dilution series, a sigmoidal binding curve is obtained. This binding curve can directly be fitted with the nonlinear solution of the law of mass action, with the dissociation constant K_D as result.

Protein Microarray

A protein microarray (or protein chip) is a high-throughput method used to track the interactions and activities of proteins, and to determine their function, and determining function on a large scale. Its main advantage lies in the fact that large numbers of proteins can be tracked in parallel. The chip consists of a support surface such as a glass slide, nitrocellulose membrane, bead, or microtitre plate, to which an array of capture proteins is bound. Probe molecules, typically labeled with a fluorescent dye, are added to the array. Any reaction between the probe and the immobilised protein emits a fluorescent signal that is read by a laser scanner. Protein microarrays are rapid, automated, economical, and highly sensitive, consuming small quantities of samples and reagents. The concept and methodology of protein microarrays was first introduced and illustrated in antibody microarrays (also referred to as antibody matrix) in 1983 in a scientific publication and a series of patents. The high-throughput technology behind the protein microarray was relatively easy to develop since it is based on the technology developed for DNA microarrays, which have become the most widely used microarrays.

Motivation for Development

Protein microarrays were developed due to the limitations of using DNA microarrays for determining gene expression levels in proteomics. The quantity of mRNA in the cell often doesn't reflect the expression levels of the proteins they correspond to. Since it is usually the protein, rather than the mRNA, that has the functional role in cell response, a novel approach was needed. Additionally post-translational modifications, which are often critical for determining protein function, are not visible on DNA microarrays. Protein microarrays replace traditional proteomics techniques such as 2D gel electrophoresis or chromatography, which were time consuming, labor-intensive and ill-suited for the analysis of low abundant proteins.

Making the Array

The proteins are arrayed onto a solid surface such as microscope slides, membranes, beads or microtitre plates. The function of this surface is to provide a support onto which proteins can be immobilized. It should demonstrate maximal binding properties, whilst maintaining the protein in its native conformation so that its binding ability is retained. Microscope slides made of glass or silicon are a popular choice since they are compatible with the easily obtained robotic arrayers and laser scanners that have been developed for DNA microarray technology. Nitrocellulose film slides are broadly accepted as the highest protein binding substrate for protein microarray applications.

The chosen solid surface is then covered with a coating that must serve the simultaneous functions of immobilising the protein, preventing its denaturation, orienting it in the appropriate direction so that its binding sites are accessible, and providing a hydrophilic environment in which the binding reaction can occur. In addition, it also needs to display minimal non-specific binding in order to minimize background noise in the detection systems. Furthermore, it needs to be compatible with different detection systems. Immobilising agents include layers of aluminium or gold, hydrophilic polymers, and polyacrylamide gels, or treatment with amines, aldehyde or epoxy. Thin-film technologies like physical vapour deposition (PVD) and chemical vapour deposition (CVD) are employed to apply the coating to the support surface.

An aqueous environment is essential at all stages of array manufacture and operation to prevent protein denaturation. Therefore, sample buffers contain a high percent of glycerol(to lower the freezing point), and the humidity of the manufacturing environment is carefully regulated. Microwells have the dual advantage of providing an aqueous environment while preventing cross-contamination between samples.

In the most common type of protein array, robots place large numbers of proteins or their ligands onto a coated solid support in a pre-defined pattern. This is known as robotic contact printing or robotic spotting. Another fabrication method is ink-jetting, a drop-on-demand, non-contact method of dispersing the protein polymers onto the solid surface in the desired pattern. Piezoelectric spotting is a similar method to ink-jet printing. The printhead moves across the array, and at each spot uses electric stimulation to deliver the protein molecules onto the surface via tiny jets. This is also a non-contact process. Photolithography is a fourth method of arraying the proteins onto the surface. Light is used in association with photomasks, opaque plates with holes or transparencies that allow light to shine through in a defined pattern. A series of chemical treatments then enables deposition of the protein in the desired pattern upon the material underneath the photomask.

The capture molecules arrayed on the solid surface may be antibodies, antigens, aptamers (nucleic acid-based ligands), affibodies (small molecules engineered to mimic monoclonal antibodies), or full length proteins. Sources of such proteins include cell-based expression systems for recombinant proteins, purification from natural sources, production in vitro by cell-free translation systems, and synthetic methods for peptides. Many of these methods can be automated for high throughput production but care must be taken to avoid conditions of synthesis or extraction that result in a denatured protein which, since it no longer recognizes its binding partner, renders the array useless.

Proteins are highly sensitive to changes in their microenvironment. This presents a challenge in maintaining protein arrays in a stable condition over extended periods of time. In situ methods involve on-chip synthesis of proteins as and when required, directly from the DNA using cell-free protein expression systems. Since DNA is a highly stable molecule it does not deteriorate over time and is therefore suited to long-term storage. This approach is also advantageous in that it circumvents the laborious and often costly processes of separate protein purification and DNA cloning, since proteins are made and immobilised simultaneously in a single step on the chip surface. Examples of In situ techniques are PISA (protein in situ array), NAPPA (nucleic acid programmable protein array) and DAPA (DNA array to protein array).

Types of Arrays

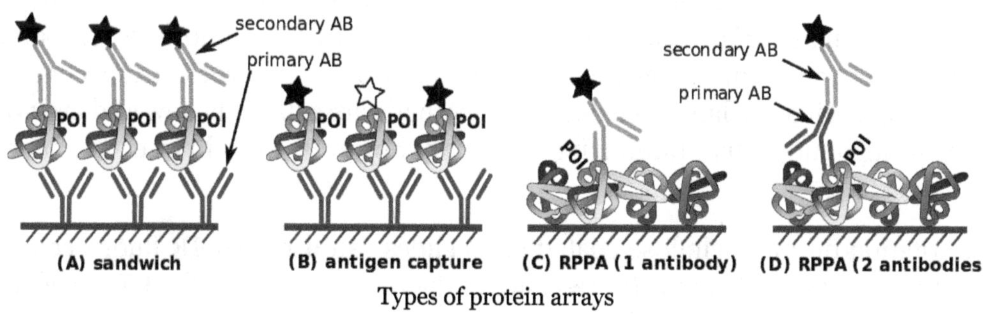

Types of protein arrays

There are three types of protein microarrays that are currently used to study the biochemical activities of proteins.

Analytical microarrays are also known as capture arrays. In this technique, a library of antibodies, aptamers or affibodies is arrayed on the support surface. These are used as capture molecules since each binds specifically to a particular protein. The array is probed with a complex protein solution such as a cell lysate. Analysis of the resulting binding reactions using various detection systems can provide information about expression levels of particular proteins in the sample as well as measurements of binding affinities and specificities. This type of microarray is especially useful in comparing protein expression in different solutions. For instance the response of the cells to a particular factor can be identified by comparing the lysates of cells treated with specific substances or grown under certain conditions with the lysates of control cells. Another application is in the identification and profiling of diseased tissues.

Functional protein microarrays (also known as target protein arrays) are constructed by immobilising large numbers of purified proteins and are used to identify protein-protein, protein-DNA, protein-RNA, protein-phospholipid, and protein-small molecule interactions, to assay enzymatic activity and to detect antibodies and demonstrate their specificity. They differ from analytical arrays in that functional protein arrays are composed of arrays containing full-length functional proteins or protein domains. These protein chips are used to study the biochemical activities of the entire proteome in a single experiment.

Reverse phase protein microarray (RPPA) involve complex samples, such as tissue lysates. Cells are isolated from various tissues of interest and are lysed. The lysate is arrayed onto the microarray and probed with antibodies against the target protein of interest. These antibodies are typically detected with chemiluminescent, fluorescent or colorimetric assays. Reference peptides are printed on the slides to allow for protein quantification of the sample lysates. RPAs allow for the determination of the presence of altered proteins or other agents that may be the result of disease. Specifically, post-translational modifications, which are typically altered as a result of disease can be detected using RPAs.

Detection

Protein array detection methods must give a high signal and a low background. The most common and widely used method for detection is fluorescence labeling which is highly sensitive, safe and compatible with readily available microarray laser scanners. Other labels can be used, such as affinity, photochemical or radioisotope tags. These labels are attached to the probe itself and can interfere with the probe-target protein reaction. Therefore, a number of label free detection methods are available, such as surface plasmon resonance (SPR), carbon nanotubes, carbon nanowire sensors (where detection occurs via changes in conductance) and microelectromechanical system (MEMS) cantilevers. All these label free detection methods are relatively new and are not yet suitable for high-throughput protein interaction detection; however, they do offer much promise for the future.

Protein quantitation on nitrocellulose coated glass slides can use near-IR fluorescent detection. This limits interferences due to auto-fluorescence of the nitrocellulose at the UV wavelengths used for standard fluorescent detection probes.

Applications

There are five major areas where protein arrays are being applied: diagnostics, proteomics, protein functional analysis, antibody characterization, and treatment development

Diagnostics involves the detection of antigens and antibodies in blood samples; the profiling of sera to discover new disease biomarkers; the monitoring of disease states and responses to therapy in personalized medicine; the monitoring of environment and food.

Proteomics pertains to protein expression profiling i.e. which proteins are expressed in the lysate of a particular cell.

Protein functional analysis is the identification of protein-protein interactions (e.g. identification of members of a protein complex), protein-phospholipid interactions, small molecule targets, enzymatic substrates (particularly the substrates of kinases) and receptor ligands.

Antibody characterization is characterizing cross-reactivity, specificity and mapping epitopes.

Treatment development involves the development of antigen-specific therapies for autoimmunity, cancer and allergies; the identification of small molecule targets that could potentially be used as new drugs.

Challenges

Despite the considerable investments made by several companies, proteins chips have yet to flood the market. Manufacturers have found that proteins are actually quite difficult to handle. A protein chip requires a lot more steps in its creation than does a DNA chip.

Challenges include: 1) finding a surface and a method of attachment that allows the proteins to maintain their secondary or tertiary structure and thus their biological activity and their interactions with other molecules, 2) producing an array with a long shelf life so that the proteins on the chip do not denature over a short time, 3) identifying and isolating antibodies or other capture molecules against every protein in the human genome, 4) quantifying the levels of bound protein while assuring sensitivity and avoiding background noise, 5) extracting the detected protein from the chip in order to further analyze it, 6) reducing non-specific binding by the capture agents, 7) the capacity of the chip must be sufficient to allow as complete a representation of the proteome to be visualized as possible; abundant proteins overwhelm the detection of less abundant proteins such as signaling molecules and receptors, which are generally of more therapeutic interest.

Two-hybrid Screening

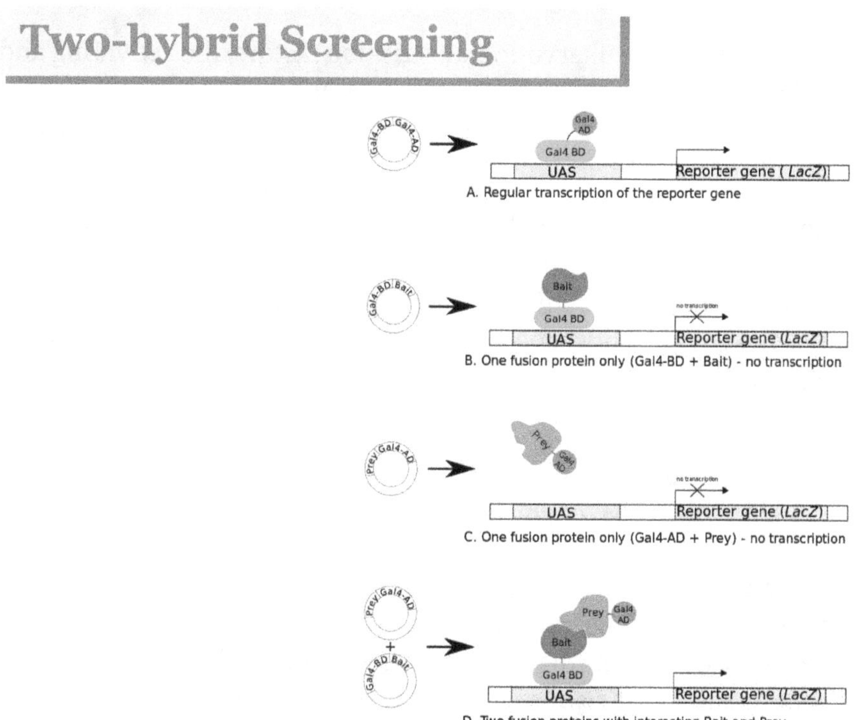

Overview of two-hybrid assay, checking for interactions between two proteins, called here *Bait* and *Prey*. A. The *Gal4* transcription factor gene produces a two-domain protein (*BD* and *AD*) essential for transcription of the reporter gene (*LacZ*). B,C. Two fusion proteins are prepared: *Gal4BD+Bait* and *Gal4AD+Prey*. Neither of them are usually sufficient to initiate transcription (of the reporter gene) alone.D. When both fusion proteins are produced and the Bait part of the first fusion protein interacts with the Prey part of the second, transcription of the reporter gene occurs.

Two-hybrid screening (also known as yeast two-hybrid system or Y2H) is a molecular biology technique used to discover protein–protein interactions (PPIs) and protein–DNA interactions by testing for physical interactions (such as binding) between two proteins or a single protein and a DNA molecule, respectively.

The premise behind the test is the activation of downstream reporter gene(s) by the binding of a transcription factor onto an upstream activating sequence (UAS). For two-hybrid screening, the transcription factor is split into two separate fragments, called the binding domain (BD) and activating domain (AD). The BD is the domain responsible for binding to the UAS and the AD is the domain responsible for the activation of transcription. The Y2H is thus a protein-fragment complementation assay.

History

Pioneered by Stanley Fields and Ok-Kyu Song in 1989, the technique was originally designed to detect protein–protein interactions using the GAL4 transcriptional activator of the yeast *Saccharomyces cerevisiae*. The GAL4 protein activated transcription of a protein involved in galactose utilization, which formed the basis of selection. Since then, the same principle has been adapted to describe many alternative methods, including some that detect protein–DNA interactions or DNA-DNA interactions, as well as methods that use *Escherichia coli* instead of yeast.

Basic Premise

The key to the two-hybrid screen is that in most eukaryotic transcription factors, the activating and binding domains are modular and can function in proximity to each other without direct binding. This means that even though the transcription factor is split into two fragments, it can still activate transcription when the two fragments are indirectly connected.

The most common screening approach is the yeast two-hybrid assay. This system often utilizes a genetically engineered strain of yeast in which the biosynthesis of certain nutrients (usually amino acids or nucleic acids) is lacking. When grown on media that lacks these nutrients, the yeast fail to survive. This mutant yeast strain can be made to incorporate foreign DNA in the form of plasmids. In yeast two-hybrid screening, separate bait and prey plasmids are simultaneously introduced into the mutant yeast strain.

Plasmids are engineered to produce a protein product in which the DNA-binding domain (BD) fragment is fused onto a protein while another plasmid is engineered to produce a protein product in which the activation domain (AD) fragment is fused onto another protein. The protein fused to the BD may be referred to as the bait protein, and is typically a known protein the investigator is using to identify new binding partners. The protein fused to the AD may be referred to as the prey protein and can be either a single known protein or a library of known or unknown proteins. In this context, a library may consist of a collection of protein-encoding sequences that represent all the proteins expressed in a particular organism or tissue, or may be generated by synthesising random DNA sequences. Regardless of the source, they are subsequently incorporated into the protein-encoding sequence of a plasmid, which is then transfected into the cells chosen for the screening method. This technique, when using a library, assumes that each cell is transfected with no more than a single plasmid and that, therefore, each cell ultimately expresses no more than a single member from the protein library.

If the bait and prey proteins interact (i.e., bind), then the AD and BD of the transcription factor are indirectly connected, bringing the AD in proximity to the transcription start site and transcription of reporter gene(s) can occur. If the two proteins do not interact, there is no transcription of the reporter gene. In this way, a successful interaction between the fused protein is linked to a change in the cell phenotype.

The challenge of separating cells that express proteins that happen to interact with their counterpart fusion proteins from those that do not, is addressed in the following section.

Fixed Domains

In any study, some of the protein domains, those under investigation, will be varied according to the goals of the study whereas other domains, those that are not themselves being investigated, will be kept constant. For example, in a two-hybrid study to select DNA-binding domains, the DNA-binding domain, BD, will be varied while the two interacting proteins, the bait and prey, must be kept constant to maintain a strong binding between the BD and AD. There are a number of domains from which to choose the BD, bait and prey and AD, if these are to remain constant. In protein–protein interaction investigations, the BD may be chosen from any of many strong DNA-binding domains such as Zif268. A frequent choice of bait and prey domains are residues 263–352 of yeast Gal11P with a N342V mutation and residues 58–97 of yeast Gal4, respectively. These domains can be used in both yeast- and bacterial-based selection techniques and are known to bind together strongly.

The AD chosen must be able to activate transcription of the reporter gene, using the cell's own transcription machinery. Thus, the variety of ADs available for use in yeast-based techniques may not be suited to use in their bacterial-based analogues. The herpes simplex virus-derived AD, VP16 and yeast Gal4 AD have been used with success in yeast whilst a portion of the α-subunit of *E. coli* RNA polymerase has been utilised in *E. coli*-based methods.

Whilst powerfully activating domains may allow greater sensitivity towards weaker interactions, conversely, a weaker AD may provide greater stringency.

Construction of Expression Plasmids

A number of engineered genetic sequences must be incorporated into the host cell to perform two-hybrid analysis or one of its derivative techniques. The considerations and methods used in the construction and delivery of these sequences differ according to the needs of the assay and the organism chosen as the experimental background.

There are two broad categories of hybrid library: random libraries and cDNA-based libraries. A cDNA library is constituted by the cDNA produced through reverse transcription of mRNA collected from specific cells of types of cell. This library can be ligated into a construct so that it is attached to the BD or AD being used in the assay. A random library uses lengths of DNA of random sequence in place of these cDNA sections. A number of methods exist for the production of these random sequences, including cassette mutagenesis. Regardless of the source of the DNA library, it is ligated into the appropriate place in the relevant plasmid/phagemid using the appropriate restriction endonucleases.

E. Coli-specific considerations

By placing the hybrid proteins under the control of IPTG-inducible *lac* promoters, they are expressed only on media supplemented with IPTG. Further, by including different antibiotic resistance genes in each genetic construct, the growth of non-transformed cells is easily prevented through culture on media containing the corresponding antibiotics. This is particularly important for counter selection methods in which a *lack* of interaction is needed for cell survival.

The reporter gene may be inserted into the *E. coli* genome by first inserting it into an episome, a type of plasmid with the ability to incorporate itself into the bacterial cell genome with a copy number of approximately one per cell.

The hybrid expression phagemids can be electroporated into *E. coli* XL-1 Blue cells which after amplification and infection with VCS-M13 helper phage, will yield a stock of library phage. These phage will each contain one single-stranded member of the phagemid library.

Recovery of Protein Information

Once the selection has been performed, the primary structure of the proteins which display the appropriate characteristics must be determined. This is achieved by retrieval of the protein-encoding sequences (as originally inserted) from the cells showing the appropriate phenotype.

E. Coli

The phagemid used to transform *E. coli* cells may be "rescued" from the selected cells by infecting them with VCS-M13 helper phage. The resulting phage particles that are produced contain the single-stranded phagemids and are used to infect XL-1 Blue cells. The double-stranded phagemids are subsequently collected from these XL-1 Blue cells, essentially reversing the process used to produce the original library phage. Finally, the DNA sequences are determined through dideoxy sequencing.

Controlling Sensitivity

The *Escherichia coli*-derived Tet-R repressor can be used in line with a conventional reporter gene and can be controlled by tetracycline or doxicycline (Tet-R inhibitors). Thus the expression of Tet-R is controlled by the standard two-hybrid system but the Tet-R in turn controls (represses) the expression of a previously mentioned reporter such as *HIS3*, through its Tet-R promoter. Tetracycline or its derivatives can then be used to regulate the sensitivity of a system utilising Tet-R.

Sensitivity may also be controlled by varying the dependency of the cells on their reporter genes. For example, this may be affected by altering the concentration of histidine in the growth medium for *his3*-dependent cells and altering the concentration of streptomycin for *aadA* dependent cells. Selection-gene-dependency may also be controlled by applying an inhibitor of the selection gene at a suitable concentration. 3-Amino-1,2,4-triazole (3-AT) for example, is a competitive inhibitor of the *HIS3*-gene product and may be used to titrate the minimum level of *HIS3* expression required for growth on histidine-deficient media.

Sensitivity may also be modulated by varying the number of operator sequences in the reporter DNA.

Non-fusion Proteins

A third, non-fusion protein may be co-expressed with two fusion proteins. Depending on the investigation, the third protein may modify one of the fusion proteins or mediate or interfere with their interaction.

Co-expression of the third protein may be necessary for modification or activation of one or both of the fusion proteins. For example, *S. cerevisiae* possesses no endogenous tyrosine kinase. If an investigation involves a protein that requires tyrosine phosphorylation, the kinase must be supplied in the form of a tyrosine kinase gene.

The non-fusion protein may mediate the interaction by binding both fusion proteins simultaneously, as in the case of ligand-dependent receptor dimerization.

For a protein with an interacting partner, its functional homology to other proteins may be assessed by supplying the third protein in non-fusion form, which then may or may not compete with the fusion-protein for its binding partner. Binding between the third protein and the other fusion protein will interrupt the formation of the reporter expression activation complex and thus reduce reporter expression, leading to the distinguishing change in phenotype.

Split-ubiquitin Yeast Two-hybrid

One limitation of classic yeast two-hybrid screens is that they are limited to soluble proteins. It is therefore impossible to use them to study the protein–protein interactions between insoluble integral membrane proteins. The split-ubiquitin system provides a method for overcoming this limitation. In the split-ubiquitin system, two integral membrane proteins to be studied are fused to two different ubiquitin moieties: a C-terminal ubiquitin moiety ("Cub", residues 35–76) and an N-terminal ubiquitin moiety ("Nub", residues 1–34). These fused proteins are called the bait and prey, respectively. In addition to being fused to an integral membrane protein, the Cub moiety is also fused to a transcription factor (TF) that can be cleaved off by ubiquitin specific proteases. Upon bait–prey interaction, Nub and Cub-moieties assemble, reconstituting the split-ubiquitin. The reconstituted split-ubiquitin molecule is recognized by ubiquitin specific proteases, which cleave off the transcription factor, allowing it to induce the transcription of reporter genes.

Fluorescent Two-hybrid Assay

Zolghadr and co-workers presented a fluorescent two-hybrid system that uses two hybrid proteins that are fused to different fluorescent proteins as well as LacI, the lac repressor. The structure of the fusion proteins looks like this: FP2-LacI-bait and FP1-prey where the bait and prey proteins interact and bring the fluorescent proteins (FP1 = GFP, FP2=mCherry) in close proximity at the binding site of the LacI protein in the host cell genome. The system can also be used to screen for inhibitors of protein–protein interactions.

Enzymatic Two-hybrid Systems: KISS

While the original Y2H system used a reconstituted transcription factor, other systems create enzymatic activities to detect PPIs. For instance, the KInase Substrate Sensor ("KISS"), is a mammalian two-hybrid approach has been designed to map intracellular PPIs. Here, a bait protein is fused to a kinase-containing portion of TYK2 and a prey is coupled to a gp130 cytokine receptor fragment. When bait and prey interact, TYK2 phosphorylates STAT3 docking sites on the prey chimera, which ultimately leads to activation of a reporter gene.

One-, Three- and One-two-hybrid Variants

One-hybrid

The one-hybrid variation of this technique is designed to investigate protein–DNA interactions and uses a single fusion protein in which the AD is linked directly to the binding domain. The binding domain in this case however is not necessarily of fixed sequence as in two-hybrid protein–protein analysis but may be constituted by a library. This library can be selected against the desired target sequence, which is inserted in the promoter region of the reporter gene construct. In a positive-selection system, a binding domain that successfully binds the UAS and allows transcription is thus selected.

Note that selection of DNA-binding domains is not necessarily performed using a one-hybrid system, but may also be performed using a two-hybrid system in which the binding domain is varied and the bait and prey proteins are kept constant.

Three-hybrid

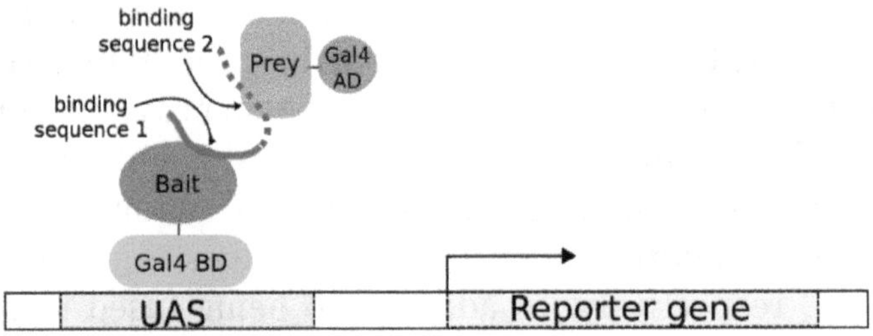

Overview of three-hybrid assay.

RNA-protein interactions have been investigated through a three-hybrid variation of the two-hybrid technique. In this case, a hybrid RNA molecule serves to adjoin together the two protein fusion domains—which are not intended to interact with each other but rather the intermediary RNA molecule (through their RNA-binding domains). Techniques involving non-fusion proteins that perform a similar function, as described in the 'non-fusion proteins' section above, may also be referred to as three-hybrid methods.

One-two-hybrid

Simultaneous use of the one- and two-hybrid methods (that is, simultaneous protein–protein and protein–DNA interaction) is known as a one-two-hybrid approach and expected to increase the stringency of the screen.

Host Organism

Although theoretically, any living cell might be used as the background to a two-hybrid analysis, there are practical considerations that dictate which is chosen. The chosen cell line should be relatively cheap and easy to culture and sufficiently robust to withstand application of the investigative methods and reagents.

Yeast

S. cerevisiae was the model organism used during the two-hybrid technique's inception. It has several characteristics that make it a robust organism to host the interaction, including the ability to form tertiary protein structures, neutral internal pH, enhanced ability to form disulfide bonds and reduced-state glutathione among other cytosolic buffer factors, to maintain a hospitable internal environment. The yeast model can be manipulated through non-molecular techniques and its complete genome sequence is known. Yeast systems are tolerant of diverse culture conditions and harsh chemicals that could not be applied to mammalian tissue cultures.

A number of yeast strains have been created specifically for Y2H screens, e.g. Y187 and AH109, both produced by Clontech. Yeast strains R2HMet and BK100 have also been used.

E. Coli

E. coli-based methods have several characteristics that may make them preferable to yeast-based homologues. The higher transformation efficiency and faster rate of growth lends *E. coli* to the use of larger libraries (in excess of 10^8). A low false positive rate of approximately $3x10^{-8}$, the absence of requirement for a nuclear localisation signal to be included in the protein sequence and the ability to study proteins that would be toxic to yeast may also be major factors to consider when choosing an experimental background organism.

It may be of note that the methylation activity of certain *E. coli* DNA methyltransferase proteins may interfere with some DNA-binding protein selections. If this is anticipated, the use of an *E. coli* strain that is defective for a particular methyltransferase may be an obvious solution.

Applications

Determination of Sequences Crucial for Interaction

the information is totally baseless. By changing specific amino acids by mutating the corresponding DNA base-pairs in the plasmids used, the importance of those amino acid residues in maintaining the interaction can be determined.

After using bacterial cell-based method to select DNA-binding proteins, it is necessary to check the specificity of these domains as there is a limit to the extent to which the bacterial cell genome can act as a sink for domains with an affinity for other sequences (or indeed, a general affinity for DNA).

Drug and Poison Discovery

Protein–protein signalling interactions pose suitable therapeutic targets due to their specificity and pervasiveness. The random drug discovery approach uses compound banks that comprise random chemical structures, and requires a high-throughput method to test these structures in their intended target.

The cell chosen for the investigation can be specifically engineered to mirror the molecular aspect that the investigator intends to study and then used to identify new human or animal therapeutics or anti-pest agents.

Determination of Protein Function

By determination of the interaction partners of unknown proteins, the possible functions of these new proteins may be inferred. This can be done using a single known protein against a library of unknown proteins or conversely, by selecting from a library of known proteins using a single protein of unknown function.

Zinc Finger Protein Selection

To select zinc finger proteins (ZFPs) for protein engineering, methods adapted from the two-hybrid screening technique have been used with success. A ZFP is itself a DNA-binding protein used in the construction of custom DNA-binding domains that bind to a desired DNA sequence.

By using a selection gene with the desired target sequence included in the UAS, and randomising the relevant amino acid sequences to produce a ZFP library, cells that host a DNA-ZFP interaction with the required characteristics can be selected. Each ZFP typically recognises only 3–4 base pairs, so to prevent recognition of sites outside the UAS, the randomised ZFP is engineered into a 'scaffold' consisting of another two ZFPs of constant sequence. The UAS is thus designed to include the target sequence of the constant scaffold in addition to the sequence for which a ZFP is selected.

A number of other DNA-binding domains may also be investigated using this system.

Strengths

- Two-hybrid screens are low-tech; they can be carried out in any lab without sophisticated equipment.
- Two-hybrid screens can provide an important first hint for the identification of interaction partners.
- The assay is scalable, which makes it possible to screen for interactions among many proteins. Furthermore, it can be automated, and by using robots many proteins can be screened against thousands of potentially interacting proteins in a relatively short time.
- Yeast two-hybrid data can be of similar quality to data generated by the alternative approach of coaffinity purification followed by mass spectrometry (AP/MS).

Weaknesses

- The main criticism applied to the yeast two-hybrid screen of protein–protein interactions is the possibility of a high number of false positive (and false negative) identifications. The exact rate of false positive results is not known, but earlier estimates were as high as 70%. The reason for this high error rate lies in the characteristics of the screen:

- Certain assay variants overexpress the fusion proteins which may cause unnatural protein concentrations that lead to unspecific (false) positives.
- The hybrid proteins are fusion proteins; that is, the fused parts may inhibit certain interactions, especially if an interaction takes place at the N-terminus of a test protein (where the DNA-binding or activation domain is typically attached).
- An interaction may not happen in yeast, the typical host organism for Y2H. For instance, if a bacterial protein is tested in yeast, it may lack a chaperone for proper folding that is only present in its bacterial host. Moreover, a mammalian protein is sometimes not correctly modified in yeast (e.g., missing phosphorylation), which can also lead to false results.
- The Y2H takes place in the nucleus. If test proteins are not localized to the nucleus (because they have other localization signals) two interacting proteins may be found to be non-interacting.
- Some proteins might specifically interact when they are co-expressed in the yeast, although in reality they are never present in the same cell at the same time. However, in most cases it cannot be ruled out that such proteins are indeed expressed in certain cells or under certain circumstances.

Each of these points alone can give rise to false results. Due to the combined effects of all error sources yeast two-hybrid have to be interpreted with caution. The probability of generating false positives means that all interactions should be confirmed by a high confidence assay, for example co-immunoprecipitation of the endogenous proteins, which is difficult for large scale protein–protein interaction data. Alternatively, Y2H data can be verified using multiple Y2H variants or bioinformatics techniques. The latter test whether interacting proteins are expressed at the same time, share some common features (such as gene ontology annotations or certain network topologies), have homologous interactions in other species.

Phage Display

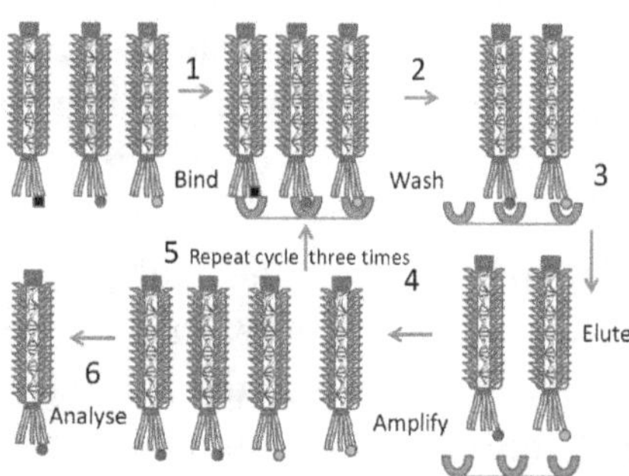

The sequence of events that are followed in phage display screening to identify polypeptides that bind with high affinity to desired target protein or DNA sequence.

Phage display is a laboratory technique for the study of protein–protein, protein–peptide, and protein–DNA interactions that uses bacteriophages (viruses that infect bacteria) to connect proteins with the genetic information that encodes them. In this technique, a gene encoding a protein of interest is inserted into a phage coat protein gene, causing the phage to "display" the protein on its outside while containing the gene for the protein on its inside, resulting in a connection between genotype and phenotype. These displaying phages can then be screened against other proteins, peptides or DNA sequences, in order to detect interaction between the displayed protein and those other molecules. In this way, large libraries of proteins can be screened and amplified in a process called *in vitro* selection, which is analogous to natural selection.

The most common bacteriophages used in phage display are M13 and fd filamentous phage, though T4, T7, and λ phage have also been used.

History

Phage display was first described by George P. Smith in 1985, when he demonstrated the display of peptides on filamentous phage by fusing the peptide of interest onto gene III of filamentous phage. A patent by George Pieczenik claiming priority from 1985 also describes the generation of phage display libraries. This technology was further developed and improved by groups at the Laboratory of Molecular Biology with Greg Winter and John McCafferty, The Scripps Research Institute with Lerner and Barbas and the German Cancer Research Center with Breitling and Dübel for display of proteins such as antibodies for therapeutic protein engineering.

Principle

Like the two-hybrid system, phage display is used for the high-throughput screening of protein interactions. In the case of M13 filamentous phage display, the DNA encoding the protein or peptide of interest is ligated into the pIII or pVIII gene, encoding either the minor or major coat protein, respectively. Multiple cloning sites are sometimes used to ensure that the fragments are inserted in all three possible reading frames so that the cDNA fragment is translated in the proper frame. The phage gene and insert DNA hybrid is then inserted (a process known as "transduction") into *Escherichia coli* (E. coli) bacterial cells such as TG1, SS320, ER2738, or XL1-Blue *E. coli*. If a "phagemid" vector is used (a simplified display construct vector) phage particles will not be released from the *E. coli* cells until they are infected with helper phage, which enables packaging of the phage DNA and assembly of the mature virions with the relevant protein fragment as part of their outer coat on either the minor (pIII) or major (pVIII) coat protein. By immobilizing a relevant DNA or protein target(s) to the surface of a microtiter plate well, a phage that displays a protein that binds to one of those targets on its surface will remain while others are removed by washing. Those that remain can be eluted, used to produce more phage (by bacterial infection with helper phage) and so produce a phage mixture that is enriched with relevant (i.e. binding) phage. The repeated cycling of these steps is referred to as 'panning', in reference to the enrichment of a sample of gold by removing undesirable materials. Phage eluted in the final step can be used to infect a suitable bacterial host, from which the phagemids can be collected and the relevant DNA sequence excised and sequenced to identify the relevant, interacting proteins or protein fragments.

The use of a helper phage can be eliminated by using 'bacterial packaging cell line' technology.

Elution can be done combining low-pH elution buffer with sonification, which, in addition to loosening the peptide-target interaction, also serves to detach the target molecule from the immobilization surface. This ultrasound-based method enables single-step selection of a high-affinity peptide.

Applications

Applications of phage display technology include determination of interaction partners of a protein (which would be used as the immobilised phage "bait" with a DNA library consisting of all coding sequences of a cell, tissue or organism) so that the function or the mechanism of the function of that protein may be determined. Phage display is also a widely used method for *in vitro* protein evolution (also called protein engineering). As such, phage display is a useful tool in drug discovery. It is used for finding new ligands (enzyme inhibitors, receptor agonists and antagonists) to target proteins. The technique is also used to determine tumour antigens (for use in diagnosis and therapeutic targeting) and in searching for protein-DNA interactions using specially-constructed DNA libraries with randomised segments.

Competing methods for *in vitro* protein evolution include yeast display, bacterial display, ribosome display, and mRNA display.

Antibody Maturation in Vitro

The invention of antibody phage display revolutionised antibody drug discovery. Initial work was done by laboratories at the MRC Laboratory of Molecular Biology (Greg Winter and John McCafferty), the Scripps Research Institute (Richard Lerner and Carlos F. Barbas) and the German Cancer Research Centre (Frank Breitling and Stefan Dübel). In 1991, The Scripps group reported the first display and selection of human antibodies on phage. This initial study described the rapid isolation of human antibody Fab that bound tetanus toxin and the method was then extended to rapidly clone human anti-HIV-1 antibodies for vaccine design and therapy.

Phage display of antibody libraries has become a powerful method for both studying the immune response as well as a method to rapidly select and evolve human antibodies for therapy. Antibody phage display was later used by Carlos F. Barbas at The Scripps Research Institute to create synthetic human antibody libraries, a principle first patented in 1990 by Breitling and coworkers (Patent CA 2035384), thereby allowing human antibodies to be created in vitro from synthetic diversity elements.

Antibody libraries displaying millions of different antibodies on phage are often used in the pharmaceutical industry to isolate highly specific therapeutic antibody leads, for development into antibody drugs primarily as anti-cancer or anti-inflammatory therapeutics. One of the most successful was HUMIRA (adalimumab), discovered by Cambridge Antibody Technology as D2E7 and developed and marketed by Abbott Laboratories. HUMIRA, an antibody to TNF alpha, was the world's first fully human antibody, which achieved annual sales exceeding $1bn.

General Protocol

Below is the sequence of events that are followed in phage display screening to identify polypeptides that bind with high affinity to desired target protein or DNA sequence:

1. Target proteins or DNA sequences are immobilised to the wells of a microtiter plate.

2. Many genetic sequences are expressed in a bacteriophage library in the form of fusions with the bacteriophage coat protein, so that they are displayed on the surface of the viral particle. The protein displayed corresponds to the genetic sequence within the phage.

3. This phage-display library is added to the dish and after allowing the phage time to bind, the dish is washed.

4. Phage-displaying proteins that interact with the target molecules remain attached to the dish, while all others are washed away.

5. Attached phage may be eluted and used to create more phage by infection of suitable bacterial hosts. The new phage constitutes an enriched mixture, containing considerably less irrelevant phage (i.e. non-binding) than were present in the initial mixture.

6. Steps 3 to 6 are optionally repeated one or more times, further enriching the phage library in binding proteins.

7. Following further bacterial-based amplification, the DNA within in the interacting phage is sequenced to identify the interacting proteins or protein fragments.

Selection of the Coat Protein

Filamentous Phages

pIII

pIII is the protein that determines the infectivity of the virion. pIII is composed of three domains (N1, N2 and CT) connected by glycine-rich linkers. The N2 domain binds to the F pilus during virion infection freeing the N1 domain which then interacts with a TolA protein on the surface of the bacterium. Insertions within this protein are usually added in position 249 (within a linker region between CT and N2), position 198 (within the N2 domain) and at the N-terminus (inserted between the N-terminal secretion sequence and the N-terminus of pIII). However, when using the BamHI site located at position 198 one must be careful of the unpaired Cysteine residue (C201) that could cause problems during phage display if one is using a non-truncated version of pIII.

An advantage of using pIII rather than pVIII is that pIII allows for monovalent display when using a phagemid (Ff-phage derived plasmid) combined with a helper phage. Moreover, pIII allows for the insertion of larger protein sequences (>100 amino acids) and is more tolerant to it than pVIII. However, using pIII as the fusion partner can lead to a decrease in phage infectivity leading to problems such as selection bias caused by difference in phage growth rate or even worse, the phage's inability to infect its host. Loss of phage infectivity can be avoided by using a phagemid plasmid and a helper phage so that the resultant phage contains both wild type and fusion pIII.

cDNA has also been analyzed using pIII via a two complementary leucine zippers system, Direct Interaction Rescue or by adding an 8-10 amino acid linker between the cDNA and pIII at the C-terminus.

pVIII

pVIII is the main coat protein of Ff phages. Peptides are usually fused to the N-terminus of pVIII.

Usually peptides that can be fused to pVIII are 6-8 amino acids long. The size restriction seems to have less to do with structural impediment caused by the added section and more to do with the size exclusion caused by pIV during coat protein export. Since there are around 2700 copies of the protein on a typical phages, it is more likely that the protein of interest will be expressed polyvalently even if a phagemid is used. This makes the use of this protein unfavorable for the discovery of high affinity binding partners.

To overcome the size problem of pVIII, artificial coat proteins have been designed. An example is Weiss and Sidhu's inverted artificial coat protein (ACP) which allows the display of large proteins at the C-terminus. The ACP's could display a protein of 20kDa, however, only at low levels (mostly only monovalently).

pVI

pVI has been widely used for the display of cDNA libraries. The display of cDNA libraries via phage display is an attractive alternative to the yeast-2-hybrid method for the discovery of interacting proteins and peptides due to its high throughput capability. pVI has been used preferentially to pVIII and pIII for the expression of cDNA libraries because one can add the protein of interest to the C-terminus of pVI without greatly affecting pVI's role in phage assembly. This means that the stop codon in the cDNA is no longer an issue. However, phage display of cDNA is always limited by the inability of most prokaryotes in producing post-translational modifications present in eukaryotic cells or by the misfolding of multi-domain proteins.

While pVI has been useful for the analysis of cDNA libraries, pIII and pVIII remain the most utilized coat proteins for phage display.

pVII and pIX

In an experiment in 1995, display of Glutathione S-transferase was attempted on both pVII and pIX and failed. However, phage display of this protein was completed successfully after the addition of a periplasmic signal sequence (pelB or ompA) on the N-terminus. In a recent study, it has been shown that AviTag, FLAG and His could be displayed on pVII without the need of a signal sequence. Then the expression of single chain Fv's (scFv), and single chain T cell receptors (scTCR) were expressed both with and without the signal sequence.

PelB (an amino acid signal sequence that targets the protein to the periplasm where a signal peptidase then cleaves off PelB) improved the phage display level when compared to pVII and pIX fusions without the signal sequence. However, this led to the incorporation of more helper phage genomes rather than phagemid genomes. In all cases, phage display levels were lower than using pIII fusion. However, lower display might be more favorable for the selection of binders due to lower display being closer to true monovalent display. In five out of six occasions, pVII and pIX fusions without pelB was more efficient than pIII fusions in affinity selection assays. The paper even goes on to state that pVII and pIX display platforms may outperform pIII in the long run.

The use of pVII and pIX instead of pIII might also be an advantage because virion rescue may be undertaken without breaking the virion-antigen bond if the pIII used is wild type. Instead, one could cleave in a section between the bead and the antigen to elute. Since the pIII is intact it does not matter whether the antigen remains bound to the phage.

T7 phages

The issue of using Ff phages for phage display is that they require the protein of interest to be translocated across the bacterial inner membrane before they are assembled into the phage. Some proteins cannot undergo this process and therefore cannot be displayed on the surface of Ff phages. In these cases, T7 phage display is used instead. In T7 phage display, the protein to be displayed is attached to the C-terminus of the gene 10 capsid protein of T7.

The disadvantage of using T7 is that the size of the protein that can be expressed on the surface is limited to shorter peptides because large changes to the T7 genome cannot be accommodated like it is in M13 where the phage just makes its coat longer to fit the larger genome within it. However, it can be useful for the production of a large protein library for scFV selection where the scFV is expressed on an M13 phage and the antigens are expressed on the surface of the T7 phage.

Bioinformatics Resources and Tools

Databases and computational tools for mimotopes have been an important part of phage display study. Databases, programs and web servers have been widely used to exclude target-unrelated peptides, characterize small molecules-protein interactions and map protein-protein interactions. Users can use three dimensional structure of a protein and the peptides selected from phage display experiment to map conformational eptiopes. Some of the fast and efficient computational methods are available online (e.g. EpiSearch http://curie.utmb.edu/episearch.html).

Tandem Affinity Purification

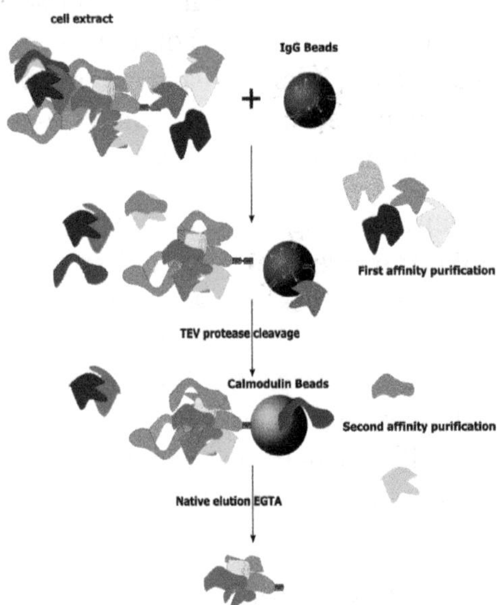

Tandem affinity purification (TAP) is a purification technique for studying protein–protein interactions. It involves creating a fusion protein with a designed piece, the TAP tag, on the end.

In the original version of the technique, the protein of interest with the TAP tag first binds to beads coated with IgG, the TAP tag is then broken apart by an enzyme, and finally a different part of the TAP tag binds reversibly to beads of a different type. After the protein of interest has been washed through two affinity columns, it can be examined for binding partners.

The original TAP method involves the fusion of the TAP tag to the C-terminus of the protein under study. The TAP tag consists of calmodulin binding peptide (CBP) from the N-terminal, followed by tobacco etch virus protease (TEV protease) cleavage site and Protein A, which binds tightly to IgG. The relative order of the modules of the tag is important because Protein A needs to be at the extreme end of the fusion protein so that the entire complex can be retrieved using an IgG matrix.

Many other tag combinations have been proposed since the TAP principle was first published.

Variant tags

This tag is also known as the C-terminal TAP tag because an N-terminal version is also available. However, the method to be described assumes the use of a C-terminal tag, although the principle behind the method is still the same.

History

TAP tagging was invented by a research team working in the European Molecular Biology Laboratory at late 1990s (Rigaut et al., 1999, Puig et al.,2001) and proposed as a new tool for proteome exploration. It was used by the team to characterize several protein complexes (Rigaut et al., 1999, Caspary et al. 1999, Bouveret et al., 2000, Puig et al., 2001). The first large-scale application of this technique was in 2002, in which the research team worked in collaboration with scientists of the proteomics company Cellzome to develop a visual map of the interaction of more than 230 multi-protein complexes in a yeast cell by systematically tagging the TAP tag to each protein. The first successful report of using TAP tag technology in plants came in 2004 (Rohila et al., 2004,)

Process

There are a few methods in which the fusion protein can be introduced into the host. If the host is yeast, then one of the methods may be the use of plasmids that will eventually translate the fusion protein within the host. Whichever method that is being used, it is preferable to maintain expression of the fusion protein as close as possible to its natural level.

Once the fusion protein is translated within the host, the new protein at one end of the fusion protein would be able to interact with other proteins. Subsequently, the fusion protein is retrieved from the host by breaking the cells and retrieving the fusion protein through affinity selection, together with the other constituents attached to the new protein, by means of an IgG matrix.

After washing, TEV protease is introduced to elute the bound material at the TEV protease cleavage site. This eluate is then incubated with calmodulin-coated beads in the presence of calcium. This second affinity step is required to remove the TEV protease as well as traces of contaminants remaining after the first affinity step. After washing, the eluate is then released with ethylene glycol tetraacetic acid (EGTA).

The native elution, consisting of the new protein and its interacting protein partners as well as CBP, can now be analyzed by sodium dodecyl sulfate polyacrylamide gel electrophoresis (SDS-PAGE) or be identified by mass spectrometry.

Advantages

An advantage of this method is that there can be real determination of protein partners quantitatively in vivo without prior knowledge of complex composition. It is also simple to execute and often provides high yield. One of the obstacles of studying protein protein interaction is the contamination of the target protein especially when we don't have any prior knowledge of it. TAP offers an effective, and highly specific means to purify target protein. After 2 successive affinity purifications, the chance for contaminants to be retained in the eluate reduces significantly.

Disadvantages

However, there is also the possibility that a tag added to a protein might obscure binding of the new protein to its interacting partners. In addition, the tag may also affect protein expression levels. On the other hand, the tag may also not be sufficiently exposed to the affinity beads, hence skewing the results.

There may also be a possibility of a cleavage of the proteins by the TEV protease, although this is unlikely to be frequent given the high specificity of the TEV protease.

Suitability

As this method involves at least 2 rounds of washing, it may not be suitable for screening transient protein interactions, unlike the yeast two-hybrid method or *in vivo* crosslinking with photo-reactive amino acid analogs. However, it is a good method for testing stable protein interactions and allows various degrees of investigation by controlling the number of times the protein complex is purified.

Applications

In 2002, the TAP tag was first used with mass spectrometry in a large-scale approach to systematically analyse the proteomics of yeast by characterizing multiprotein complexes. The study revealed 491 complexes, 257 of them wholly new. The rest were familiar from other research, but now virtually all of them were found to have new components. They drew up a map relating all the protein components functionally in a complex network.

Many other proteomic analyses also involve the use of TAP tag. A research by EMBO (Dziembowski, 2004) identified a new complex required for nuclear pre-mRNA retention and splicing. They have purified a novel trimeric complex composed of 3 other subunits (Snu17p, Bud13p and Pml1p) and find that these subunits are not essential for viability but required for efficient splicing (removal of introns) of pre-mRNA. In 2006, *Fleischer et al.* systematically identified proteins associated with eukaryotic ribosomal complexes. They used multifaceted mass spectrometry proteomic screens to identify yeast ribosomal complexes and then used TAP tagging to functionally link up all these proteins.

Other Epitope-tag Combinations

The principle of tandem-affinity purification of multiprotein complexes is not limited to the combination of CBP and Protein A tags used in the original work by Rigaut et al. (1999). For example, the combination of FLAG- and HA-tags has been used since 2000 by the group of Nakatani to purify numerous protein complexes from mammalian cells. Many other tag combinations have been proposed since the TAP principle was published.

Bimolecular Fluorescence Complementation

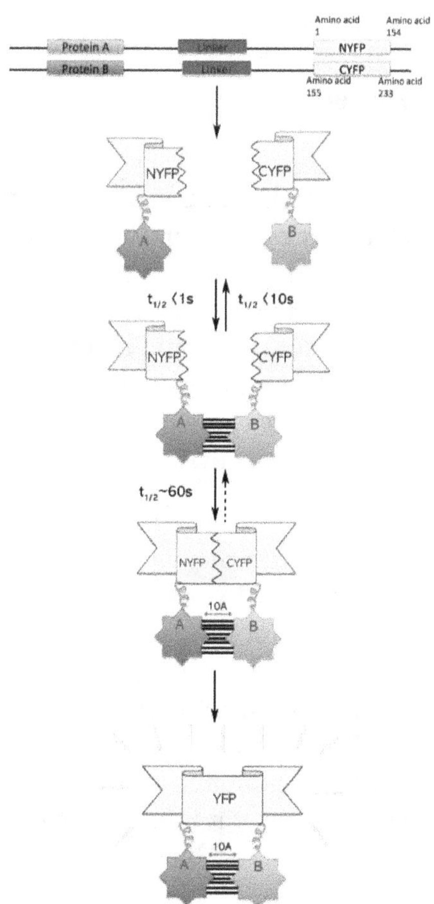

Protein complex formation using BiFC. Interaction between protein A and protein B occurs first, followed by the reformation and fluorescence of fluorescent reporter protein

Bimolecular fluorescence complementation (also known as BiFC) is a technology typically used to validate protein interactions. It is based on the association of fluorescent protein fragments that are attached to components of the same macromolecular complex. Proteins that are postulated to interact are fused to unfolded complementary fragments of a fluorescent reporter protein and expressed in live cells. Interaction of these proteins will bring the fluorescent fragments within proximity, allowing the reporter protein to reform in its native three-dimensional structure and emit its fluorescent signal. This fluorescent signal can be detected and located within the cell using an inverted fluorescence microscope that allows imaging of fluorescence in cells. In addition, the intensity of the fluorescence emitted is proportional to the strength of the interaction, with stronger levels of fluorescence indicating close or direct interactions and lower fluorescence levels suggesting interaction within a complex. Therefore, through the visualisation and analysis of the intensity and distribution of fluorescence in these cells, one can identify both the location and interaction partners of proteins of interest.

History

Biochemical complementation was first reported in subtilisin-cleaved bovine pancreatic ribonuclease, then expanded using β-galactosidase mutants that allowed cells to grow on lactose.

Recognition of many proteins' ability to spontaneously assemble into functional complexes as well as the ability of protein fragments to assemble as a consequence of the spontaneous functional complex assembly of interaction partners to which they are fused was later reported for ubiquitin fragments in yeast protein interactions.

In 2000, Ghosh *et al* developed a system that allowed a green fluorescent protein (GFP) to be reassembled using an anti-parallel leucine zipper in *E. coli* cells. This was achieved by dissecting GFP into C- and N-terminal GFP fragments. As the GFP fragment was attached to each leucine zipper by a linker, the heterodimerisation of the anti-parallel leucine zipper resulted in a reconstituted, or re-formed, GFP protein that could be visualised. The successful fluorescent signal indicated that the separate GFP peptide fragments were able to correctly reassemble and achieve tertiary folding. It was, therefore, postulated that using this technique, fragmented GFP could be used to study interaction of protein–protein pairs that have their N–C termini in close proximity.

After the demonstration of successful fluorescent protein fragment reconstitution in mammalian cells, Hu *et al.* described the use of fragmented yellow fluorescent protein (YFP) in the investigation of bZIP and Rel family transcription factor interactions. This was the first report bZIP protein interaction regulation by regions outside of the bZIP domain, regulation of subnuclear localization of the bZIP domains Fos and Jun by their different interacting partners, and modulation of transcriptional activation of bZIP and Rel proteins through mutual interactions. In addition, this study was the first report of an *in vivo* technique, now known as the bimolecular fluorescence complementation (BiFC) assay, to provide insight into the structural basis of protein complex formation through detection of fluorescence caused by the assembly of fluorescent reporter protein fragments tethered to interacting proteins.

Fluorescent Labeling

Fluorophore activation occurs through an autocatalytic cyclization reaction that occurs after the protein has been folded correctly. This was advanced with the successful reconstitution of the YFP fluorophore from protein fragments that had been fused to interacting proteins within 8 hours of transfection was reported in 2002.

Workflow

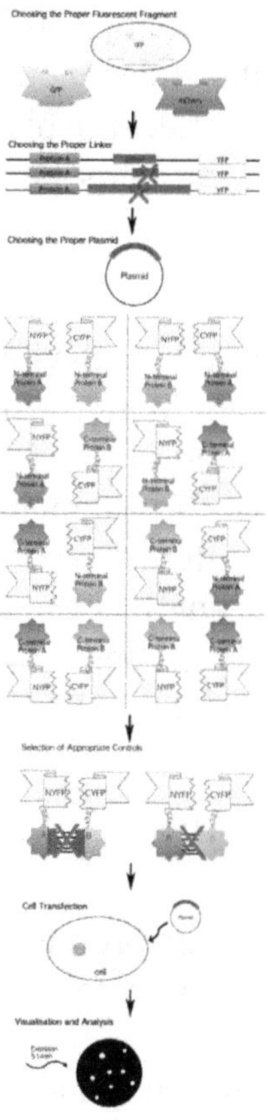

Wo4rkflow for BiFC

Selection of Fusion Protein Production System

There are different production systems that can be used for the fusion protein generated. Transient gene expression is used to identify protein–protein interactions *in vivo* as well as in subcellular localisation of the BiFC complex. However, one must be cautious against protein over-expression, as this may skew both preferential localisation and the predominant protein complexes formed. Instead, weak promoters, the use of low levels of plasmid DNA in the transfection, and plasmid vectors that do not replicate in mammalian cells should be used to express proteins at or near their endogenous levels to mimic the physiological cellular environment. Also, careful selection of the fluorescent protein is important, as different fluorescent proteins require different cellular environments. For example, GFP can be used in *E. coli* cells, while YFP is used in mammalian cells.

Stable cell lines with the expression vector integrated into its genome allows more stable gene expression in the cell population, resulting in more consistent results.

Determination of Fusion Sites

When deciding the linker fusion site on the protein surface, there are three main considerations. First, the fluorescent protein fragments must be able to associate with one another when their tethered proteins interact. Structural information and the location of the interaction surface may be useful when determining the fusion site to the linker, although the information is not necessary, as multiple combinations and permutations can be screened. Secondly, the creation of the fusion protein must not significantly alter the localisation, stability, or expression of the proteins to which the fragments are linked as compared to the endogenous wild-type proteins. Finally, the addition of the fluorescent fragment fusion must not affect the biological function of the protein, preferably verified using assays that evaluate all of the proteins' known functions.

Designing Linkers

A linker is a short amino acid sequence that tethers the fluorescent reporter protein fragment to the protein of interest, forming the fusion protein. When designing a linker sequence, one must ensure that the linker is sufficiently soluble and long to provide the fluorescent protein fragments with flexibility and freedom of movement so that the fragment and its partner fragment will collide frequently enough to reconstitute during the interaction of their respective fused proteins. Although it is not documented, it is possible that the length or the sequence of the linker may influence complementation of some proteins. Reported linker sequences RSIAT and RPACKIPNDLKQKVMNH (single amino acid code) and AAANSSIDLISVPVDSR (Sigma) have been successfully used in BiFC experiments.

Creating Proper Plasmid Expression Vectors

When designing plasmid vectors to express the proteins of interest, the construct must be able to express proteins that are able to form fusion proteins with fluorescent protein fragments without disrupting the protein's function. In addition, the expected protein complex must be able to accept stabilisation of the fluorescent protein fragment interaction without affecting the protein complex function or the cell being studied. Many fluorescent protein fragments that combine in several ways can be used in BiFC. Generally, YFP is recommended to serve as the reporter protein, cleaved at residue 155 (N-terminal consisting of residues 1–154 and C-terminal consisting of residues 155–238) or residue 173 in particular, as these sets of fragments are highly efficient in their complementation when fused to many interacting proteins and they produce low levels fluorescence when fused to non-interacting proteins. It is suggested that each target protein is fused to both the N- and C-terminal fragments of the fluorescent reporter protein in turn, and that the fragments are fused at each of the N- and C-terminal ends of the target proteins. This will allow a total of eight different permutations, with interactions being tested:

N-terminal fragment fused at the N-terminal protein 1 + C-terminal fragment fused at the N-terminal protein 2N-terminal fragment fused at the N-terminal protein 1 + C-terminal fragment fused at the C-terminal protein 2N-terminal fragment fused at the C-terminal protein 1 + C-terminal fragment fused at the N-terminal protein 2N-terminal fragment fused at the C-terminal protein 1 + C-terminal fragment fused at the C-terminal protein 2C-terminal fragment fused at the N-terminal protein 1 + N-terminal fragment fused at the N-terminal protein 2C-terminal fragment fused at the N-terminal protein 1 + N-terminal fragment fused at the C-terminal protein 2C-terminal fragment fused at the C-terminal protein 1 + N-terminal fragment fused at the N-terminal protein 2C-terminal fragment fused at the C-terminal protein 1 + N-terminal fragment fused at the C-terminal protein 2

Selection of Appropriate Cell Culture System

As previously stated, it is important to ensure that the fluorescent reporter protein being used in

BiFC is appropriate and can be expressed in the cell culture system of choice, as not all reporter proteins can fluoresce or be visualised in all model systems.

Selection of Appropriate Controls

Fluorescent protein fragments can associate and fluoresce at low efficiency in the absence of a specific interaction. Therefore, it is important to include controls to ensure that the fluorescence from fluorescent reporter protein reconstitution is not due to unspecific contact.

Some controls include fluorophore fragments linked to non-interacting proteins, as the presence of these fusions tend to decrease non-specific complementation and false positive results.

Another control is created by linking the fluorescent protein fragment to proteins with mutated interaction faces. So long as the fluorescent fragment is fused to the mutated proteins in the same manner as the wild-type protein, and the gene expression levels and localisation are unaffected by the mutation, this serves as a strong negative control, as the mutant proteins, and therefore, the fluorescent fragments, should be unable to interact.

Internal controls are also necessary to normalise for differences in transfection efficiencies and gene expression in different cells. This is accomplished by co-transfecting cells with plasmids encoding the fusion proteins of interest as well as a whole (non-fragmented) protein that fluoresces at a different wavelength from the fluorescent reporter protein. During visualisation, one determines the fluorescence intensities of the BiFC complex and the internal control which, after subtracting background signal, becomes a ratio. This ratio represents the BiFC efficiency and can be compared with other ratios to determine the relative efficiencies of the formation of different complexes.

Cell Transfection

Once the fusion proteins and controls have been designed and generated in their appropriate expression system, the plasmids must be transfected into the cells to be studied. After transfection, one must wait, typically about eight hours, to allow time for the fusion proteins to interact and their linked fluorescent reporter protein fragments to associate and fluoresce.

Visualisation and Analysis

After sufficient time for the fusion proteins and their linked fluorescent fragments to interact and fluoresce, the cells can be observed under an inverted fluorescence microscope that can visualise fluorescence in cells. Although the fluorescence intensity of BiFC complexes is usually <10% of that produced by expression of intact fluorescent proteins, the extremely low autofluorescence in the visible range extremely most cells often makes the BiFC signal orders of magnitude higher than background fluorescence.

If fluorescence is detected when the fusion proteins are expressed, but is lacking or significantly reduced after the expression of the mutated negative control, it is likely that a specific interaction occurs between the two target proteins of interest. However, if the fluorescence intensity is not significantly different between the mutated negative control fusion protein and its wild-type counterpart, then the fluorescence is likely caused by non-specific protein interactions, so a different combination of fusion protein conformations should be tested.

If no fluorescence is detected, an interaction may still exist between the proteins of interest, as the creation of the fusion protein may alter the structure or interaction face of the target protein or the fluorescence fragments may be physically unable to associate. To ensure that this result is not a false negative, that there is no interaction, the protein interaction must be tested in a situation where fluorescence complementation and activation requires an external signal. In this case, if the external signal fails to cause fluorescence fragment association, it is likely that the proteins do not interact or there is a physical impediment to fluorescence complementation.

Strengths

Relevant Biological Context

Proteins interact with different protein partners and other macromolecules to achieve functions that support different functions in cells that support survival of the organism. Identifying these interactions may provide clues to their effects on cell processes. As these interactions can be affected by both the internal environment and external stimuli, studying these interactions *in vivo* and at endogenous levels, as is recommended in BiFC, provides a physiologically-relevant context from which to draw conclusions about protein interactions.

Direct Visualisation

BiFC enables direct visualisation of protein interactions in living cells with limited cell perturbation, rather than relying on secondary effects or staining by exogenous molecules that can fail to distribute evenly. This, and the ability to observe the living cells for long periods of time, is made possible by the strong intrinsic fluorescence of the reconstituted reporter protein reduces the chances of an incorrect readout associated with the protein isolation process.

Sensitivity

Unlike many *in vivo* protein-interaction assays, BiFC does not require protein complexes to be formed by a large proportion of the proteins or at stoichiometric proportions. Instead, BiFC can detect interactions among protein subpopulations, weak interactions, and low expression proteins due to the stable complementation of the fluorescent reporter protein. In addition, successful fluorescent protein reconstitution has been reported for protein partners over 7 nm apart, so long as the linkers binding the fluorophore fragment to the protein of interest has the flexibility needed to associate with its corresponding fragment. Furthermore, the strength of the protein interaction can be quantitatively determined by changes in fluorescent signal strength.

Spatial Resolution

BiFC allows measurement of spatial and temporal changes in protein complexes, even in response to activating and inhibiting drugs and subcellularly, providing the highest spatial resolution of *in vivo* protein–protein interaction assays.

No specialised Equipment

BiFC does not require specialised equipment, as visualisation is possible with an inverted fluorescence microscope that can detect fluorescence in cells. In addition, analysis does not require complex data processing or correction for other sources of fluorescence.

No structural Information Needed

BiFC can be performed without structural information about the interaction partners, so long as the fluorescent reporter protein fragments can associate within the complex, as multiple combinations of fusion proteins can be screened. This is due to the assumption that, since the protein functions are recapitulated in the *in vivo* context, the complex structure will resemble that of the intact proteins seen physiologically.

Multiple Applications

The BiFC technology has been refined and expanded to include the abilities to simultaneously visualise multiple protein complexes in the same cell, RNA/protein interactions, to quickly detect changes in gene transduction pathways, demonstrate hidden phenotypes of drugs, where the predicted treatment outcome (i.e. cell death, differentiation, morphological change) is not seen *in vivo*, study complex formation in different cellular compartments, and to map protein interaction surfaces

Limitations

Real-time Detection

The fluorescent signal only is produced after the proteins have interacted, which is generally in the order of hours. Hence BiFC is unable to provide real-time detection of protein interactions. The delay for chemical reactions to generate fluorophore may also have an effect on the dynamics of complex dissociation and partner exchange.

Irreversible BiFC Formation

BiFC complex formation is only reversible during the initial step of fluorescent reporter protein re-assembly, typically in the order of milliseconds. Once the fluorochrome has been reconstituted, it is essentially irreversible *in vitro*. This prevents proteins from interacting with others and may disrupt the association/disassociation of protein complexes in dynamic equilibrium.

Independent Fluorescent Protein Fragment Associations

Fluorescent protein fragments have a limited ability to associate independent of the proteins to which they are fused. Although protein-independent association will vary depending on identities of the fusion proteins and their expression levels, one must provide the necessary and numerous controls to distinguish between true and false-positive protein interactions. Generally, this limitation is mitigated by ensuring that the fusion proteins of interest are expressed at endogenous concentrations.

Altering Protein Structure and Steric Hindrance

Fluorescent fragment linkage may alter the folding or structure of the protein of interest, leading to the elimination of an interacting protein's surface binding site. In addition, the arrangement of the fluorescent fragments may prevent fluorophore reconstitution through steric hindrance, although steric hindrance can be reduced or eliminated by using a linker sequence that allows sufficient flexibility for the fluorescent fragments to associate. Therefore, absence of fluorescence complementation may be a false negative and does not necessarily prove that the interaction in question does not occur.

Obligate Anaerobes

Due to the requirement of molecular oxygen for fluorophore formation, BiFC cannot be used in obligate anaerobes, which cannot survive in the presence of oxygen. This limits the use of BiFC to aerobic organisms.

Use of Fusion Proteins

Because endogenous wild-type proteins cannot be visualised *in vivo*, fusion proteins must be created and their plasmids transfected into the cells studied. These fusion proteins may not recapitulate the functions, localisation, and interactions common to their wild-type counterparts, providing an inaccurate picture of the proteins in question. This problem can be alleviated by using structural information and the location of interaction sites to rationally identify fusion sites on the proteins of interest, using appropriate controls, and comparing the expression levels and functions of the fusion and wild-type proteins through Western Blots and functional assays.

Temperature Dependence

Although low temperatures favour the reconstitution of fluorescence when fragments are within proximity, this may affect the behaviour of the target proteins leading to inaccurate conclusions regarding the nature of protein interactions and their interacting partners.

Exact Interaction Relationship Unknown

Because fluorophore reconstitution can occur at a distance of 7 nm or more, fluorescence complementation may indicate either a direct or indirect (i.e. within the same complex) interaction between the fluorescent fragments' fused proteins.

Application

In addition to the validation of protein–protein interactions described above, BiFC has been expanded and adapted to other applications:

Multicolour Fluorescence

The fluorescent protein fragments used in BiFC have been expanded to include the colours blue, cyan, green, yellow, red, cherry, and Venus. This range in colours has made the development of multicolour fluorescence complementation analysis possible. This technique allows multiple protein complexes to be visualised simultaneously in the same cell. In addition, proteins typically have a large number of alternative interaction partners. Therefore, by fusing fragments of different fluorescent proteins to candidate proteins, one can study competition between alternative interaction partners for complex formation through the complementation of different fluorescent colour fragments.

RNA-binding Protein Interactions

BiFC has been expanded to include the study of RNA-binding protein interactions in a method Rackham and Brown described as trimolecular fluorescence complementation (TriFC). In this method, a fragment of the Venus fluorescent protein is fused to the mRNA of interest, and the complementary Venus portion fused to the RNA-binding protein of interest. Similar to BiFC, if the mRNA and protein interact, the Venus protein will be reconstituted and fluoresce. Also known as the RNA bridge method, as the fluorophore and other interacting proteins form a bridge between the protein and the RNA of interest, this allows a simple detection and localisation of RNA-protein interactions within a living cell and provides a simple method to detect direct or indirect RNA-protein association (i.e. within a complex) that can be verified through in vitro analysis of purified compounds or RNAi knockdown of the bridging molecule(s).

Pathway Organisation and Signal Transduction Cascades

BiFC can be used to link genes to one other and their function through measurement of interactions among the proteins that the genes encode. This application is ideal for novel genes in which little is known about their up- and downstream effectors, as novel pathway linkages can be made. In addition, the effects of drugs, hormones, or deletion or knockdown of the gene of interest, and the subsequent effects on both the strength of the protein–protein interactions and the location of the interaction can be observed within seconds.

Complex Formation in Different Cellular Compartments

BiFC has been used to study nuclear translocation, via complex localisation, as well as interactions involving integral membrane proteins. Thus, BiFC is an important tool in understanding transcription factor localisation in subcellular compartments.

Quantifying Protein–protein Interaction Surfaces

BiFC has been coupled with flow cytometry (BiFC-FC). This allows protein–protein interaction surfaces to be mapped through the introduction of site-directed or random mutations that affect complex formation.

Comparisons to Other Technologies

Most techniques used to study protein–protein interactions rely on *in vitro* methods. Unfortunately, studying proteins in an artificial system, outside of their cellular environment, poses a number of difficulties. For example, this may require the removal of proteins from their normal cellular environment. The processing required to isolate the protein may affect its interactions with other proteins. In addition, isolating the protein from the intracellular signaling and mechanisms that occur in the normal cell may provide a misleading picture of intracellular and physiological occurrences. Furthermore, proteins studied in vitro may be studied at concentrations vastly different from their normal abundance levels, may not necessarily be transported efficiently into the cells, or may not be selective enough to function in the host genome. Finally, by studying proteins *in vitro*, one is unable to determine the influence of specific protein–protein interactions in the cell on the functional or physiological consequences.

Other *in vivo* assays most commonly used to study protein–protein interactions include fluorescence resonance energy transfer (FRET) and yeast two-hybrid (Y2H) assay. Each of these assays has their advantages and disadvantages in comparison to BiFC:

Fluorescence Resonance Energy Transfer (FRET)

Fluorescence resonance energy transfer (FRET), also known as förster resonance energy transfer, resonance energy transfer (RET) or electronic energy transfer (EET), is based on the transfer of energy from an excited (donor) chromophore or fluorophore (if the chromophores are fluorescent) to a nearby acceptor. In this method, fluorophores are chemically linked or genetically fused to two proteins hypothesised to interact. If the proteins interact, this will bring the fluorophores into close spatial proximity. If the fluorophores are oriented in a manner that exposes the fluorophores to one another, usually ensured when designing and constructing the fluorophore-protein linkage/ fusion, then the energy transfer from the excited donor fluorophore will result in a change in the fluorescent intensities or lifetimes of the fluorophores.

Yeast Two-hybrid (Y2H)

The yeast two-hybrid (Y2H) is a genetic screening technique that can be used to detect physical (binding) protein–protein or protein–DNA interactions. It tests a 'bait' protein of known function that is fused to the binding domain of the transcription factor GAL4 against potential interacting proteins or a cDNA library that express the GAL4 activation domain (the 'prey').

Technology Comparisons

Comparison Technology	Similarity to BiFC	Advantages	Disadvantages	
FRET	Ability to detect and locate protein interaction sites within live cells	**Instantaneous real-time monitoring of protein interactions** This allows the detection of proteins with short-lived associations. **Reversible fluorophore interaction** More complex interaction dynamics can be detected (i.e. dynamic equilibrium – continuous complex formation and dissociation).	**Close spatial proximity** Potential interacting proteins must be close to one another, typically within 60–100Å, for the energy transfer to occur. In contrast, reconstitution of the BiFC fluorophore is possible at a distance of over 7nm and only requires that the link between the fluorophore fragment and the protein of interest be sufficiently flexible to allow association. **Decreased sensitivity** BiFC is typically more sensitive due to its stable protein complementation. In contrast, FRET produces background fluorescence when the acceptor fluorophore is excited. Therefore, numerous controls must be performed to quantify the change in fluorescence intensity in the presence versus the absence of energy transfer between fluorophores. Consequently, it is difficult to detect weak interactions using FRET . FRET usually requires higher levels of gene expression to detect the energy transfer, as the proportion of proteins that form complexes must be high enough to produce a detectable change in donor and acceptor fluorescence intensities. **Irreversible photo-bleaching** FRET quantification requires irreversible photobleaching or fluorescence lifetime imaging that uses instrumentation that is not widely available. This complicates the visualisation process, as the fluorescence will be destroyed over time by the light needed to excite the fluorophores.	FRET: interaction between protein A and protein B brings the two fluorescent proteins together and energy transfer occurs between the two fluorescent proteins

Y2H	*In vivo* technique used to screen for interactions	**Genetic interaction screen** Y2H provides an *in vivo* technique for studying transcription factor binding at the sequence level in eukaryotic cells	**Tentative bait-prey linkage** The link between the bait and prey is often tentative and often ambiguous, as the study of over-expressed fusion proteins in the yeast nucleus can give rise to non-specific interactions, and incompatible systems (i.e. mammalian proteins do not always express correctly in yeast cells, proteins may co-express unnaturally). **Erroneous transcription activation** The reporter gene can easily be transcriptionally activated, providing a false positive readout of protein–protein or protein–DNA interaction. **Genetic complementation** Different alleles of the same gene can result in genetic complementation, leading to inaccurate protein–DNA interactions. **Overexpression of proteins** Y2H requires proteins to be overexpressed, which can skew data analysis, as proteins may interact with different proteins or DNA when expressed at higher-than-normal levels as compared to the interactions that occur at endogenous gene expression levels. **Nuclear localisation** Y2H can only be used in the nucleus, where transcriptional activation occurs. This both removes the protein from its physiologically relevant environment, disregarding the role of the protein's normal environment on protein interactions, and limits detection to proteins that are active in the nucleus.	Yeast-2-Hybrid: interaction between protein A and protein B activates transcription

Chromatin Immunoprecipitation

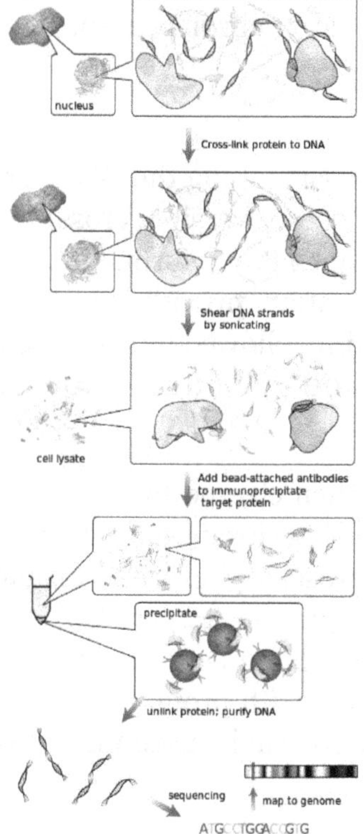

ChIP-sequencing workflow

Chromatin Immunoprecipitation (ChIP) is a type of immunoprecipitation experimental technique used to investigate the interaction between proteins and DNA in the cell. It aims to determine whether specific proteins are associated with specific genomic regions, such as transcription factors on promoters or other DNA binding sites, and possibly defining cistromes. ChIP also aims to determine the specific location in the genome that various histone modifications are associated with, indicating the target of the histone modifiers.

Briefly, the conventional method is as follows:

1. DNA and associated proteins on chromatin in living cells or tissues are crosslinked (this step is omitted in Native ChIP).
2. The DNA-protein complexes (chromatin-protein) are then sheared into ~500 bp DNA fragments by sonication or nuclease digestion.
3. Cross-linked DNA fragments associated with the protein(s) of interest are selectively immunoprecipitated from the cell debris using an appropriate protein-specific antibody.
4. The associated DNA fragments are purified and their sequence is determined. Enrichment of specific DNA sequences represents regions on the genome that the protein of interest is associated with *in vivo*.

Typical ChIP

There are mainly two types of ChIP, primarily differing in the starting chromatin preparation. The first uses reversibly cross-linked chromatin sheared by sonication called cross-linked ChIP (XChIP). Native ChIP (NChIP) uses native chromatin sheared by micrococcal nuclease digestion.

Cross-linked ChIP (XChIP)

Cross-linked ChIP is mainly suited for mapping the DNA target of transcription factors or other chromatin-associated proteins, and uses reversibly cross-linked chromatin as starting material. The agent for reversible cross-linking could be formaldehyde or UV light. Then the cross-linked chromatin is usually sheared by sonication, providing fragments of 300 - 1000 base pairs (bp) in length. Mild formaldehyde crosslinking followed by nuclease digestion has been used to shear the chromatin. Chromatin fragments of 400 - 500bp have proven to be suitable for ChIP assays as they cover two to three nucleosomes.

Cell debris in the sheared lysate is then cleared by sedimentation and protein–DNA complexes are selectively immunoprecipitated using specific antibodies to the protein(s) of interest. The antibodies are commonly coupled to agarose, sepharose or magnetic beads. The immunoprecipitated complexes (i.e., the bead–antibody–protein–target DNA sequence complex) are then collected and washed to remove non-specifically bound chromatin, the protein–DNA cross-link is reversed and proteins are removed by digestion with proteinase K. An epitope-tagged version of the protein of interest, or *in vivo* biotinylation can be used instead of antibodies to the native protein of interest.

The DNA associated with the complex is then purified and identified by polymerase chain reaction

(PCR), microarrays (ChIP-on-chip), molecular cloning and sequencing, or direct high-throughput sequencing (ChIP-Seq).

Native ChIP (NChIP)

Native ChIP is mainly suited for mapping the DNA target of histone modifiers. Generally, native chromatin is used as starting chromatin. As histones wrap around DNA to form nucleosomes, they are naturally linked. Then the chromatin is sheared by micrococcal nuclease digestion, which cuts DNA at the length of the linker, leaving nucleosomes intact and providing DNA fragments of one nucleosome (200bp) to five nucleosomes (1000bp) in length.

Thereafter, methods similar to XChIP are used for clearing the cell debris, immunoprecipitating the protein of interest, removing protein from the immunoprecipated complex, and purifying and analyzing the complex-associated DNA.

Comparison of XChIP and NChIP

The major advantage for NChIP is antibody specificity. It is important to note that most antibodies to modified histones are raised against unfixed, synthetic peptide antigens and that the epitopes they need to recognize in the XChIP may be disrupted or destroyed by formaldehyde cross-linking, particularly as the cross-links are likely to involve lysine e-amino groups in the N-terminals, disrupting the epitopes. This is likely to explain the consistently low efficiency of XChIP protocols compare to NChIP.

But XChIP and NChIP have different aims and advantages relative to each other. XChIP is for mapping target sites of transcription factors and other chromatin associated proteins; NChIP is for mapping target sites of histone modifiers.

Table 1 Advantages and Disadvantages of NChIP and XChIP

	XChIP	**NChIP**
Advantages	Suitable for transcriptional factors, or any other weakly binding chromatin associated proteins. Applicable to any organisms where native protein is hard to prepare	Testable antibody specificity Better antibody specificity as target protein naturally intact Better chromatin and protein recovery efficiency due to Better antibody specificity
Disadvantages	Inefficient chromatin recovery due to antibody target protein epitope disruption May cause false positive result due to fixation of transient proteins to chromatin Wide range of chromatin shearing size due to random cut by sonication.	Usually not suitable for non-histone proteins Nucleosomes may rearrange during digestion

History and New ChIP methods

In 1984 John T. Lis and David Gilmour, at the time a graduate student in the Lis lab, used UV irradiation, a zero-length protein-nucleic acid crosslinking agent, to covalently cross-link proteins bound to DNA in living bacterial cells. Following lysis of cross-linked cells and immunoprecipitation of bacterial RNA polymerase, DNA associated with enriched RNA polymerase was hybridized to probes corresponding to different regions of known genes to determine the in vivo distribution and density of RNA polymerase at these genes. A year later they used the same methodology to study distribution of eukaryotic RNA polymerase II on fruit fly heat shock genes. These reports are considered the pioneering studies in the field of chromatin immunoprecipitation. XChIP was further modified and developed by Alexander Varshavsky and co-workers, who examined distribution of histone H4 on heat shock genes using formaldehyde cross-linking. This technique was extensively developed and refined thereafter. NChIP approach was first described by Hebbes *et al.*, 1988, and also been developed and refined quickly. The typical ChIP assay usually take 4–5 days, and require 10^6~ 10^7 cells at least. Now new techniques on ChIP could be achieved as few as 100~1000 cells and complete within one day.

- Carrier ChIP (CChIP): This approach could use as few as 100 cells by adding *Drosophila* cells as carrier chromatin to reduce loss and facilitate precipitation of the target chromatin. However, it demands highly specific primers for detection of the target cell chromatin from the foreign carrier chromatin background, and it takes two to three days.
- Fast ChIP (qChIP): The fast ChIP assay reduced the time by shortening two steps in a typical ChIP assay: *(i)* an ultrasonic bath accelerates the rate of antibody binding to target proteins—and thereby reduces immunoprecipitation time *(ii)* a resin-based (Chelex-100) DNA isolation procedure reduces the time of cross-link reversal and DNA isolation. However, the fast protocol is suitable only for large cell samples (in the range of 10^6~10^7). Up to 24 sheared chromatin samples can be processed to yield PCR-ready DNA in 5 hours, allowing multiple chromatin factors be probed simultaneously and/or looking at genomic events over several time points.
- Quick and quantitative ChIP (Q^2ChIP) : The assay uses 100,000 cells as starting material and is suitable for up to 1,000 histone ChIPs or 100 transcription factor ChIPs. Thus many chromatin samples can be prepared in parallel and stored, and Q^2ChIP can be undertaken in a day.
- MicroChIP (μChIP): chromatin is usually prepared from 1,000 cells and up to 8 ChIPs can be done in parallel without carriers. The assay can also start with 100 cells, but only suit for one ChIP. It can also use small (1 mm^3) tissue biopsies and microChIP can be done within one day.
- Matrix ChIP: This is a microplate-based ChIP assay with increased throughput and simplified the procedure. All steps are done in microplate wells without sample transfers, enabling a potential for automation. It enables 96 ChIP assays for histone and various DNA-bound proteins in a single day.

ChIP has also been applied for genome wide analysis by combining with microarray technology (ChIP-on-chip) or second generation DNA-sequencing technology (Chip-Sequencing). ChIP can also combine with paired-end tags sequencing in Chromatin Interaction Analysis using Paired End Tag sequencing (ChIA-PET), a technique developed for large-scale, de novo analysis of higher-order chromatin structures.

Limitations of ChIP

- Large Scale assays using ChIP is challenging using intact model organisms. This is because antibodies have to be generated for each TF, or, alternatively, transgenic model organisms expressing epitope-tagged TFs need to be produced.
- Researchers studying differential gene expression patterns in small organisms also face problems as genes expressed at low levels, in a small number of cells, in narrow time window.
- ChIP experiments cannot discriminate between different TF isoforms (Protein isoform).

References

- Kevin Downard (24 August 2007). Mass Spectrometry of Protein Interactions. John Wiley & Sons. ISBN 978-0-470-14632-3. Retrieved 14 September 2013.
- Scott JS, Barbas CF III, Burton, DA (2001). Phage Display: A Laboratory Manual. Plainview, N.Y: Cold Spring Harbor Laboratory Press. ISBN 0-87969-740-7.
- Lowman HB, Clackson T (2004). "1.3". Phage display: a practical approach. Oxford [Oxfordshire]: Oxford University Press. pp. 10–11. ISBN 0-19-963873-X.
- "DNA Microarrays: Techniques". Arabidopsis.info. Archived from the original on August 28, 2008. Retrieved 2013-01-19.

6

Interactive Network of Proteins

In molecular biology, the interactions that occur in a particular cell is known as an interactome whereas RNA interference is the process in which RNA molecules inhibit gene expression. This section elucidates the crucial aspects of the interactive network of proteins.

Interactome

Part of the DISC1 interactome with genes represented by text in boxes and interactions noted by lines between the genes. From Hennah and Porteous, 2009.

In molecular biology, an interactome is the whole set of molecular interactions in a particular cell. The term specifically refers to physical interactions among molecules (such as those among proteins, also known as protein–protein interactions) but can also describe sets of indirect interactions among genes (genetic interactions). Mathematically, interactomes are generally displayed as graphs.

The word "interactome" was originally coined in 1999 by a group of French scientists headed by Bernard Jacq. Though interactomes may be described as biological networks, they should not be confused with other networks such as neural networks or food webs.

Molecular Interaction Networks

Molecular interactions can occur between molecules belonging to different biochemical families (proteins, nucleic acids, lipids, carbohydrates, etc.) and also within a given family. Whenever such molecules are connected by physical interactions, they form molecular interaction networks that are generally classified by the nature of the compounds involved. Most commonly, *interactome* refers to *protein–protein interaction* (PPI) network (PIN) or subsets thereof. For instance, the Sirt-1 protein interactome and Sirt family second order interactome is the network involving Sirt-1 and its directly interacting proteins where as second order interactome illustrates interactions up to second order of neighbors (Neighbors of neighbors). Another extensively studied type of interactome is the protein–DNA interactome, also called a *gene-regulatory network*, a network formed by transcription factors, chromatin regulatory proteins, and their target genes. Even *metabolic networks* can be considered as molecular interaction networks: metabolites, i.e. chemical compounds in a cell, are converted into each other by enzymes, which have to bind their substrates physically.

In fact, all interactome types are interconnected. For instance, protein interactomes contain many enzymes which in turn form biochemical networks. Similarly, gene regulatory networks overlap substantially with protein interaction networks and signaling networks.

Size of Interactomes

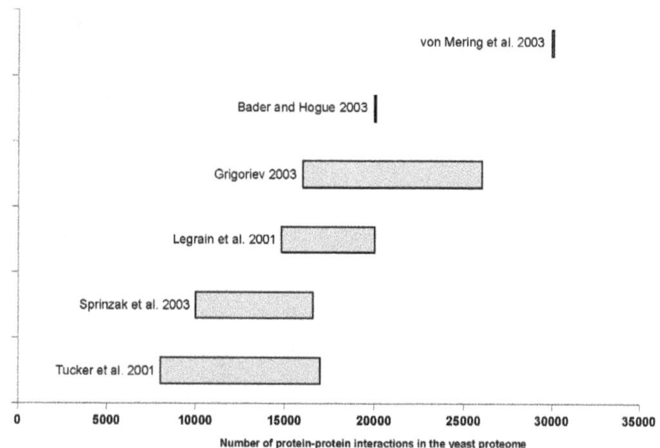

Estimates of the yeast protein interactome. From

It has been suggested that the size of an organism's interactome correlates better than genome size with the biological complexity of the organism. Although protein–protein interaction maps containing several thousand binary interactions are now available for several species, none of them is presently complete and the size of interactomes is still a matter of debate.

Yeast

The yeast interactome, i.e. all protein–protein interactions among proteins of *Saccharomyces cerevisiae*, has been estimated to contain between 10,000 and 30,000 interactions. A reasonable estimate may be on the order of 20,000 interactions. Larger estimates often include indirect or predicted interactions, often from affinity purification/mass spectrometry (AP/MS) studies.

Genetic Interaction Networks

Genes interact in the sense that they affect each other's function. For instance, a mutation may be harmless, but when it is combined with another mutation, the combination may turn out to be lethal. Such genes are said to "interact genetically". Genes that are connected in such a way form *genetic interaction networks*. Some of the goals of these networks are: develop a functional map of a cell's processes, drug target identification, and to predict the function of uncharacterized genes.

In 2010, the most "complete" gene interactome produced to date was compiled from about 5.4 million two-gene comparisons to describe "the interaction profiles for ~75% of all genes in the budding yeast," with ~170,000 gene interactions. The genes were grouped based on similar function so as to build a functional map of the cell's processes. Using this method the study was able to predict known gene functions better than any other genome-scale data set as well as adding functional information for genes that hadn't been previously described. From this model genetic interactions can be observed at multiple scales which will assist in the study of concepts such as gene conservation. Some of the observations made from this study are that there were twice as many negative as positive interactions, negative interactions were more informative than positive interactions, and genes with more connections were more likely to result in lethality when disrupted.

Interactomics

Interactomics is a discipline at the intersection of bioinformatics and biology that deals with studying both the interactions and the consequences of those interactions between and among proteins, and other molecules within a cell. Interactomics thus aims to compare such networks of interactions (i.e., interactomes) between and within species in order to find how the traits of such networks are either preserved or varied.

Interactomics is an example of "top-down" systems biology, which takes an overhead, as well as overall, view of a biosystem or organism. Large sets of genome-wide and proteomic data are collected, and correlations between different molecules are inferred. From the data new hypotheses are formulated about feedbacks between these molecules. These hypotheses can then be tested by new experiments.

Experimental Methods to Map Interactomes

The study of interactomes is called interactomics. The basic unit of a protein network is the protein–protein interaction (PPI). While there are numerous methods to study PPIs, there are relatively few that have been used on a large scale to map whole interactomes.

The yeast two hybrid system (Y2H) is suited to explore the binary interactions among two proteins at a time. Affinity purification and subsequent mass spectrometry is suited to identify a protein complex. Both methods can be used in a high-throughput (HTP) fashion. Yeast two hybrid screens allow include false positive interactions between proteins that are never expressed in the same time and place; affinity capture mass spectrometry does not have this drawback, and is the current gold standard. Yeast two-hybrid data better indicates non-specific tendencies towards sticky interactions rather while affinity capture mass spectrometry better indicates functional in vivo protein–protein interactions.

Computational Methods to Study Interactomes

Once an interactome has been created, there are numerous ways to analyze its properties. However, there are two important goals of such analyses. First, scientists try to elucidate the systems properties of interactomes, e.g. the topology of its interactions. Second, studies may focus on individual proteins and their role in the network. Such analyses are mainly carried out using bioinformatics methods and include the following, among many others:

Validation

First, the coverage and quality of an interactome has to be evaluated. Interactomes are never complete, given the limitations of experimental methods. For instance, it has been estimated that typical Y2H screens detect only 25% or so of all interactions in an interactome. The coverage of an interactome can be assessed by comparing it to benchmarks of well-known interactions that have been found and validated by independent assays.Other methods filter out false positives calculating the simliarity of known annotations of the proteins involved or define a likelyhood of interaction using the subcellular localization of these proteins.

Predicting PPIs

Using experimental data as a starting point, *homology transfer* is one way to predict interactomes. Here, PPIs from one organism are used to predict interactions among homologous proteins in another organism ("*interologs*"). However, this approach has certain limitations, primarily because the source data may not be reliable (e.g. contain false positives and false negatives). In addition, proteins and their interactions change during evolution and thus may have been lost or gained. Nevertheless, numerous interactomes have been predicted, e.g. that of *Bacillus licheniformis*.

Some algorithms use experimental evidence on structural complexes, the atomic details of binding interfaces and produce detailed atomic models of protein–protein complexes as well as other protein–molecule interactions. Other algorithms use only sequence information, thereby creating unbiased complete networks of interaction with many mistakes.

Some methods use machine learning to distinguish how interacting protein pairs differ from non-interacting protein pairs in terms of pairwise features such as cellular colocalization, gene co-expression, how closely located on a DNA are the genes that encode the two proteins, and so on. Random Forest has been found to be most-effective machine learning method for protein interaction prediction. Such methods have been applied for discovering protein interactions on human interactome, specifically the interactome of Membrane proteins and the interactome of Schizophrenia-associated proteins.

Text Mining of PPIs

Some efforts have been made to extract systematically interaction networks directly from the scientific literature. Such approaches range in terms of complexity from simple co-occurrence statistics of entities that are mentioned together in the same context (e.g. sentence) to sophisticated natural language processing and machine learning methods for detecting interaction relationships.

Protein Function Prediction

Protein interaction networks have been used to predict the function of proteins of unknown functions. This is usually based on the assumption that uncharacterized proteins have similar functions as their interacting proteins (*guilt by association*). For example, YbeB, a protein of unknown function was found to interact with ribosomal proteins and later shown to be involved in translation. Although such predictions may be based on single interactions, usually several interactions are found. Thus, the whole network of interactions can be used to predict protein functions, given that certain functions are usually enriched among the interactors.

Perturbations and Disease

The *topology* of an interactome makes certain predictions how a network reacts to the perturbation (e.g. removal) of nodes (proteins) or edges (interactions). Such perturbations can be caused by mutations of genes, and thus their proteins, and a network reaction can manifest as a disease. A network analysis can identified drug targets and biomarkers of diseases.

Network Structure and Topology

Interaction networks can be analyzed using the tools of graph theory. Network properties include the degree distribution, clustering coefficients, betweenness centrality, and many others. The distribution of properties among the proteins of an interactome has revealed that the interactome networks often have scale-free topology where functional modules within a network indicate specialized subnetworks. Such modules can be functional, as in a signaling pathway, or structural, as in a protein complex. In fact, it is a formidable task to identify protein complexes in an interactome, given that a network on its own does not directly reveal the presence of a stable complex.

Studied Interactomes

Viral Interactomes

Viral protein interactomes consist of interactions among viral or phage proteins. They were among the first interactome projects as their genomes are small and all proteins can be analyzed with limited resources. Viral interactomes are connected to their host interactomes, forming virus-host interaction networks. Some published virus interactomes include

Bacteriophage

- *Escherichia coli* bacteriophage lambda
- *Escherichia coli* bacteriophage T7
- *Streptococcus pneumoniae* bacteriophage Dp-1
- *Streptococcus pneumoniae* bacteriophage Cp-1

The lambda and VZV interactomes are not only relevant for the biology of these viruses but also for technical reasons: they were the first interactomes that were mapped with multiple Y2H vectors, proving an improved strategy to investigate interactomes more completely than previous attempts have shown.

Human (Mammalian) Viruses

- Human varicella zoster virus (VZV)
- Chandipura virus
- Epstein-Barr virus (EBV)
- Hepatitis C virus (HPC), Human-HCV interactions
- Hepatitis E virus (HEV)
- Herpes simplex virus 1 (HSV-1)
- Kaposi's sarcoma-associated herpesvirus (KSHV)
- Murine cytomegalovirus (mCMV)

Bacterial Interactomes

Relatively few bacteria have been comprehensively studied for their protein–protein interactions. However, none of these interactomes are complete in the sense that they captured all interactions. In fact, it has been estimated that none of them covers more than 20% or 30% of all interactions, primarily because most of these studies have only employed a single method, all of which discover only a subset of interactions. Among the published bacterial interactomes (including partial ones) are

Species	proteins total	interactions	type
Helicobacter pylori	1,553	~3,004	Y2H
Campylobacter jejuni	1,623	11,687	Y2H
Treponema pallidum	1,040	3,649	Y2H
Escherichia coli	4,288	(5,993)	AP/MS
Escherichia coli	4,288	2,234	Y2H
Mesorhizobium loti	6,752	3,121	Y2H
Mycobacterium tuberculosis	3,959	>8000	B2H
Mycoplasma genitalium	482		AP/MS
Synechocystis sp. PCC6803	3,264	3,236	Y2H
Staphylococcus aureus (MRSA)	2,656	13,219	AP/MS

The *E. coli* and *Mycoplasma* interactomes have been analyzed using large-scale protein complex affinity purification and mass spectrometry (AP/MS), hence it is not easily possible to infer direct interactions. The others have used extensive yeast two-hybrid (Y2H) screens. The *Mycobacterium tuberculosis* interactome has been analyzed using a bacterial two-hybrid screen (B2H).

Note that numerous additional interactomes have been predicted using computational methods.

Eukaryotic Interactomes

There have been several efforts to map eukaryotic interactomes through HTP methods. While no biological interactomes have been fully characterized, over 90% of proteins in *Saccharomyces cerevisiae* have been screened and their interactions characterized, making it the best-characterized interactome. Species whose interactomes have been studied in some detail include

- *Schizosaccharomyces pombe*
- *Caenorhabditis elegans*

- *Drosophila melanogaster*
- *Homo sapiens*

Recently, the pathogen-host interactomes of Hepatitis C Virus/Human (2008), Epstein Barr virus/Human (2008), Influenza virus/Human (2009) were delineated through HTP to identify essential molecular components for pathogens and for their host's immune system.

Predicted Interactomes

As described above, PPIs and thus whole interactomes can be predicted. While the reliability of these predictions is debatable, they are providing hypotheses that can be tested experimentally. Interactomes have been predicted for a number of species, e.g.

- Human (*Homo sapiens*)
- Rice (*Oryza sativa*)
- *Xanthomonas oryzae*
- *Arabidopsis thaliana*
- Tomato
- *Brassica rapa*
- *Maize, corn (Zea mays)*
- *Populus trichocarpa*

Network Properties of Interactomes

Protein interaction networks can be analyzed with the same tool as other networks. In fact, they share many properties with biological or social networks. Some of the main characteristics are as follows.

The *Treponema pallidum* protein interactome.

Degree Distribution

The degree distribution describes the number of proteins that have a certain number of connections. Most protein interaction networks show a scale-free (power law) degree distribution where the connectivity distribution $P(k) \sim k^{-\gamma}$ with k being the degree. This relationship can also be seen as a straight line on a log-log plot since, the above equation is equal to $\log(P(k)) \sim -\gamma \cdot \log(k)$. One characteristic of such distributions is that there are many proteins with few interactions and few proteins that have many interactions, the latter being called "hubs".

Hubs

Highly connected nodes (proteins) are called hubs. Han et al. have coined the term "party hub" for hubs whose expression is correlated with its interaction partners. Party hubs also connect proteins within functional modules such as protein complexes. In contrast, "date hubs" do not exhibit such a correlation and appear to connect different functional modules. Party hubs are found predominantly in AP/MS data sets, whereas date hubs are found predominantly in binary interactome network maps. Note that the validity of the date hub/party hub distinction was disputed. Party hubs generally consist of multi-interface proteins whereas date hubs are more frequently single-interaction interface proteins. Consistent with a role for date-hubs in connecting different processes, in yeast the number of binary interactions of a given protein is correlated to the number of phenotypes observed for the corresponding mutant gene in different physiological conditions.

Modules

Nodes involved in the same biochemical process are highly interconnected.

Interactome Evolution

The evolution of interactome complexity is delineated in a study published in Nature. In this study it is first noted that the boundaries between prokaryotes, unicellular eukaryotes and multicellular eukaryotes are accompanied by orders-of-magnitude reductions in effective population size, with concurrent amplifications of the effects of random genetic drift. The resultant decline in the efficiency of selection seems to be sufficient to influence a wide range of attributes at the genomic level in a nonadaptive manner. The Nature study shows that the variation in the power of random genetic drift is also capable of influencing phylogenetic diversity at the subcellular and cellular levels. Thus, population size would have to be considered as a potential determinant of the mechanistic pathways underlying long-term phenotypic evolution. In the study it is further shown that a phylogenetically broad inverse relation exists between the power of drift and the structural integrity of protein subunits. Thus, the accumulation of mildly deleterious mutations in populations of small size induces secondary selection for protein–protein interactions that stabilize key gene functions, mitigating the structural degradation promoted by inefficient selection. By this means, the complex protein architectures and interactions essential to the genesis of phenotypic diversity may initially emerge by non-adaptive mechanisms.

Criticisms, Challenges, and Responses

Kiemer and Cesareni raise the following concerns with the state (circa 2007) of the field especially with the comparative interactomic: The experimental procedures associated with the field are error prone leading to "noisy results". This leads to 30% of all reported interactions being artifacts. In fact, two groups using the same techniques on the same organism found less than 30% interactions in common. However, some authors have argued that such non-reproducibility results from the extraordinary sensitivity of various methods to small experimental variation. For instance, identical conditions in Y2H assays result in very different interactions when different Y2H vectors are used.

Techniques may be biased, i.e. the technique determines which interactions are found. In fact, any method has built in biases, especially protein methods. Because every protein is different no method can capture the properties of each protein. For instance, most analytical methods that work fine with soluble proteins deal poorly with membrane proteins. This is also true for Y2H and AP/MS technologies.

Interactomes are not nearly complete with perhaps the exception of *S. cerevisiae*. This is not really a criticism as any scientific area is "incomplete" initially until the methodologies have been improved. Interactomics in 2015 is where genome sequencing was in the late 1990s, given that only a few interactome datasets are available.

While genomes are stable, interactomes may vary between tissues, cell types, and developmental stages. Again, this is not a criticism, but rather a description of the challenges in the field.

It is difficult to match evolutionarily related proteins in distantly related species. While homologous DNA sequences can be found relatively easily, it is much more difficult to predict homologous interactions ("interologs") because the homologs of two interacting proteins do not need to interact. For instance, even within a proteome two proteins may interact but their paralogs may not.

Each protein–protein interactome may represent only a partial sample of potential interactions, even when a supposedly definitive version is published in a scientific journal. Additional factors may have roles in protein interactions that have yet to be incorporated in interactomes. The binding strength of the various protein interactors, microenvironmental factors, sensitivity to various procedures, and the physiological state of the cell all impact protein–protein interactions, yet are usually not accounted for in interactome studies.

RNA Interference

RNA interference (RNAi) is a biological process in which RNA molecules inhibit gene expression, typically by causing the destruction of specific mRNA molecules. Historically, it was known by other names, including *co-suppression, post-transcriptional gene silencing* (PTGS), and *quelling*. Only after these apparently unrelated processes were fully understood did it become clear that they all described the RNAi phenomenon. Andrew Fire and Craig C. Mello shared the 2006 Nobel Prize in Physiology or Medicine for their work on RNA interference in the nematode worm *Caenorhabditis elegans*, which they published in 1998. Since the discovery of RNAi and its regulatory potentials, it has become evident that RNAi has immense potential in suppression of desired genes. RNAi is now known as precise, efficient, stable and better than antisense technology for gene suppression.

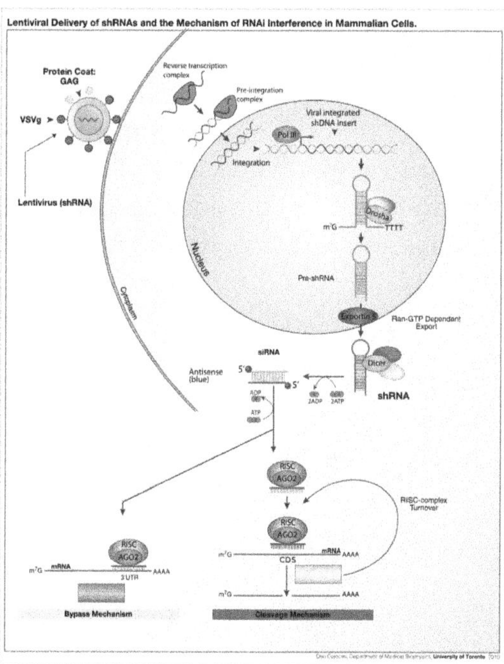

Lentiviral delivery of designed shRNA's and the mechanism of RNA interference in mammalian cells.

Two types of small ribonucleic acid (RNA) molecules – microRNA (miRNA) and small interfering RNA (siRNA) – are central to RNA interference. RNAs are the direct products of genes, and these small RNAs can bind to other specific messenger RNA (mRNA) molecules and either increase or decrease their activity, for example by preventing an mRNA from producing a protein. RNA interference has an important role in defending cells against parasitic nucleotide sequences – viruses and transposons. It also influences development.

The RNAi pathway is found in many eukaryotes, including animals, and is initiated by the enzyme Dicer, which cleaves long double-stranded RNA (dsRNA) molecules into short double-stranded fragments of ~20 nucleotide siRNAs. Each siRNA is unwound into two single-stranded RNAs (ssRNAs), the passenger strand and the guide strand. The passenger strand is degraded and the guide strand is incorporated into the RNA-induced silencing complex (RISC). The most well-studied outcome is post-transcriptional gene silencing, which occurs when the guide strand pairs with a complementary sequence in a messenger RNA molecule and induces cleavage by Argonaute, the catalytic component of the RISC complex. In some organisms, this process spreads systemically, despite the initially limited molar concentrations of siRNA.

RNAi is a valuable research tool, both in cell culture and in living organisms, because synthetic dsRNA introduced into cells can selectively and robustly induce suppression of specific genes of interest. RNAi may be used for large-scale screens that systematically shut down each gene in the cell, which can help to identify the components necessary for a particular cellular process or an event such as cell division. The pathway is also used as a practical tool in biotechnology, medicine and insecticides.

Cellular Mechanism

RNAi is an RNA-dependent gene silencing process that is controlled by the RNA-induced silencing complex (RISC) and is initiated by short double-stranded RNA molecules in a cell's cytoplasm, where they interact with the catalytic RISC component argonaute. When the dsRNA is exogenous (coming from infection by a virus with an RNA genome or laboratory manipulations), the RNA is imported directly into the cytoplasm and cleaved to short fragments by Dicer. The initiating dsRNA can also be endogenous (originating in the cell), as in pre-microRNAs expressed from RNA-coding genes in the genome. The primary transcripts from such genes are first processed to form the characteristic stem-loop structure of pre-miRNA in the nucleus, then exported to the cytoplasm. Thus, the two dsRNA pathways, exogenous and endogenous, converge at the RISC.

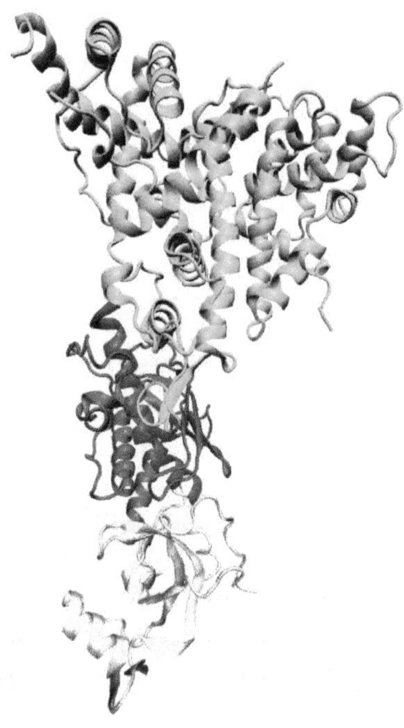

The dicer protein from *Giardia intestinalis*, which catalyzes the cleavage of dsRNA to siRNAs. The RNase domains are colored green, the PAZ domain yellow, the platform domain red, and the connector helix blue.

DsRNA Cleavage

Endogenous dsRNA initiates RNAi by activating the ribonuclease protein Dicer, which binds and cleaves double-stranded RNAs (dsRNAs) to produce double-stranded fragments of 20–25 base pairs with a 2-nucleotide overhang at the 3' end. Bioinformatics studies on the genomes of multiple organisms suggest this length maximizes target-gene specificity and minimizes non-specific effects. These short double-stranded fragments are called small interfering RNAs (siRNAs). These siRNAs are then separated into single strands and integrated into an active RISC complex. After integration into the RISC, siRNAs base-pair to their target mRNA and cleave it, thereby preventing it from being used as a translation template.

Exogenous dsRNA is detected and bound by an effector protein, known as RDE-4 in *C. elegans* and R2D2 in *Drosophila*, that stimulates dicer activity. This protein only binds long dsRNAs, but the mechanism producing this length specificity is unknown. This RNA-binding protein then facilitates the transfer of cleaved siRNAs to the RISC complex.

In *C. elegans* this initiation response is amplified through the synthesis of a population of 'secondary' siRNAs during which the dicer-produced initiating or 'primary' siRNAs are used as templates.

These 'secondary' siRNAs are structurally distinct from dicer-produced siRNAs and appear to be produced by an RNA-dependent RNA polymerase (RdRP).

MicroRNA

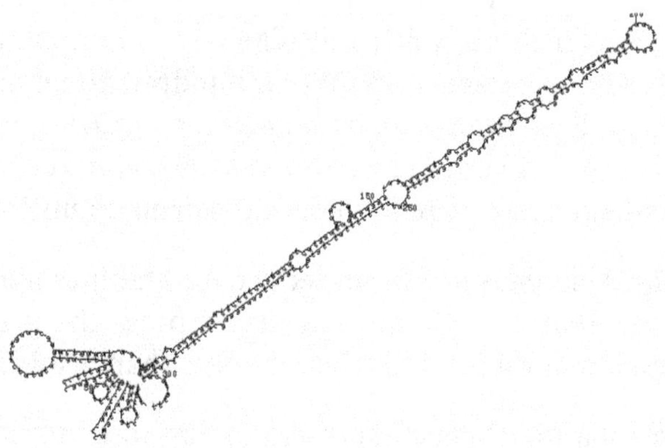

The stem-loop secondary structure of a pre-microRNA from *Brassica oleracea*.

MicroRNAs (miRNAs) are genomically encoded non-coding RNAs that help regulate gene expression, particularly during development. The phenomenon of RNA interference, broadly defined, includes the endogenously induced gene silencing effects of miRNAs as well as silencing triggered by foreign dsRNA. Mature miRNAs are structurally similar to siRNAs produced from exogenous dsRNA, but before reaching maturity, miRNAs must first undergo extensive post-transcriptional modification. A miRNA is expressed from a much longer RNA-coding gene as a primary transcript known as a *pri-miRNA* which is processed, in the cell nucleus, to a 70-nucleotide stem-loop structure called a *pre-miRNA* by the microprocessor complex. This complex consists of an RNase III enzyme called Drosha and a dsRNA-binding protein DGCR8. The dsRNA portion of this pre-miRNA is bound and cleaved by Dicer to produce the mature miRNA molecule that can be integrated into the RISC complex; thus, miRNA and siRNA share the same downstream cellular machinery. First, viral encoded miRNA was described in EBV. Thereafter, an increasing number of microRNAs have been described in viruses. VIRmiRNA is a comprehensive catalogue covering viral microRNA, their targets and anti-viral miRNAs.

siRNAs derived from long dsRNA precursors differ from miRNAs in that miRNAs, especially those in animals, typically have incomplete base pairing to a target and inhibit the translation of many different mRNAs with similar sequences. In contrast, siRNAs typically base-pair perfectly and induce mRNA cleavage only in a single, specific target. In *Drosophila* and *C. elegans*, miRNA and siRNA are processed by distinct argonaute proteins and dicer enzymes.

Three Prime Untranslated Regions and MicroRNAs

Three prime untranslated regions (3'UTRs) of messenger RNAs (mRNAs) often contain regulatory sequences that post-transcriptionally cause RNA interference. Such 3'-UTRs often contain both binding sites for microRNAs (miRNAs) as well as for regulatory proteins. By binding to specific sites within the 3'-UTR, miRNAs can decrease gene expression of various mRNAs by either inhibiting translation or directly causing degradation of the transcript. The 3'-UTR also may have silencer regions that bind repressor proteins that inhibit the expression of a mRNA.

The 3'-UTR often contains microRNA response elements (MREs). MREs are sequences to which miRNAs bind. These are prevalent motifs within 3'-UTRs. Among all regulatory motifs within the 3'-UTRs (e.g. including silencer regions), MREs make up about half of the motifs.

As of 2014, the miRBase web site, an archive of miRNA sequences and annotations, listed 28,645 entries in 233 biologic species. Of these, 1,881 miRNAs were in annotated human miRNA loci. miRNAs were predicted to have an average of about four hundred target mRNAs (affecting expression of several hundred genes). Friedman et al. estimate that >45,000 miRNA target sites within human mRNA 3'UTRs are conserved above background levels, and >60% of human protein-coding genes have been under selective pressure to maintain pairing to miRNAs.

Direct experiments show that a single miRNA can reduce the stability of hundreds of unique mRNAs. Other experiments show that a single miRNA may repress the production of hundreds of proteins, but that this repression often is relatively mild (less than 2-fold).

The effects of miRNA dysregulation of gene expression seem to be important in cancer. For instance, in gastrointestinal cancers, nine miRNAs have been identified as epigenetically altered and effective in down regulating DNA repair enzymes.

The effects of miRNA dysregulation of gene expression also seem to be important in neuropsychiatric disorders, such as schizophrenia, bipolar disorder, major depression, Parkinson's disease, Alzheimer's disease and autism spectrum disorders.

RISC Activation and Catalysis

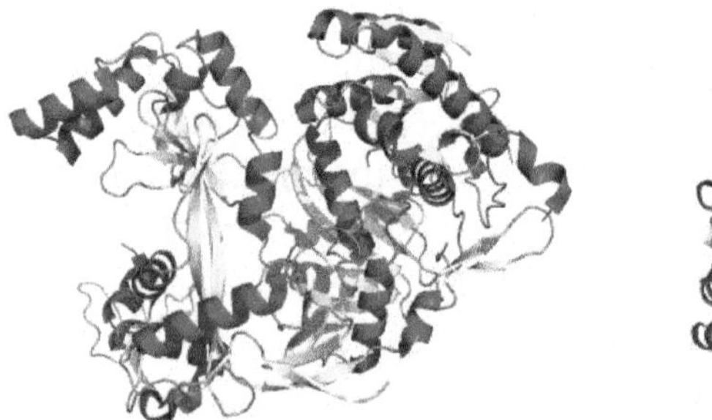

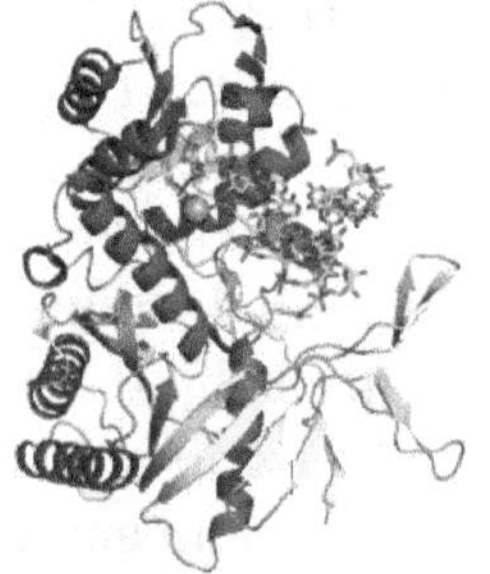

Left: A full-length argonaute protein from the archaea species *Pyrococcus furiosus*. *Right:* The PIWI domain of an argonaute protein in complex with double-stranded RNA.

The active components of an RNA-induced silencing complex (RISC) are endonucleases called argonaute proteins, which cleave the target mRNA strand complementary to their bound siRNA. As the fragments produced by dicer are double-stranded, they could each in theory produce a functional siRNA. However, only one of the two strands, which is known as the *guide strand*, binds the argonaute protein and directs gene silencing. The other *anti-guide strand* or *passenger strand* is degraded during RISC activation. Although it was first believed that an ATP-dependent helicase separated these two strands, the process proved to be ATP-independent and performed directly by the protein components of RISC. However, an *in vitro* kinetic analysis of RNAi in the presence and absence of ATP showed that ATP may be required to unwind and remove the cleaved mRNA strand from the RISC complex after catalysis. The guide strand tends to be the one whose 5' end is less stably paired to its complement, but strand selection is unaffected by the direction in which dicer cleaves the dsRNA before RISC incorporation. Instead, the R2D2 protein may serve as the differentiating factor by binding the more-stable 5' end of the passenger strand.

The structural basis for binding of RNA to the argonaute protein was examined by X-ray crystallography of the binding domain of an RNA-bound argonaute protein. Here, the phosphorylated 5' end of the RNA strand enters a conserved basic surface pocket and makes contacts through a divalent cation (an atom with two positive charges) such as magnesium and by aromatic stacking (a process that allows more than one atom to share an electron by passing it back and forth) between the 5' nucleotide in the siRNA and a conserved tyrosine residue. This site is thought to form a nucleation site for the binding of the siRNA to its mRNA target. Analysis of the inhibitory effect of mismatches in either the 5' or 3' end of the guide strand has demonstrated that the 5' end of the guide strand is likely responsible for matching and binding the target mRNA, while the 3' end is responsible for physically arranging target mRNA into a cleavage-favorable RISC region.

It is not understood how the activated RISC complex locates complementary mRNAs within the cell. Although the cleavage process has been proposed to be linked to translation, translation of the mRNA target is not essential for RNAi-mediated degradation. Indeed, RNAi may be more effective against mRNA targets that are not translated. Argonaute proteins are localized to specific regions in the cytoplasm called P-bodies (also cytoplasmic bodies or GW bodies), which are regions with high rates of mRNA decay; miRNA activity is also clustered in P-bodies. Disruption of P-bodies decreases the efficiency of RNA interference, suggesting that they are a critical site in the RNAi process.

Transcriptional Silencing

Components of the RNAi pathway are used in many eukaryotes in the maintenance of the organization and structure of their genomes. Modification of histones and associated induction of heterochromatin formation serves to downregulate genes pre-transcriptionally; this process is referred to as RNA-induced transcriptional silencing (RITS), and is carried out by a complex of proteins called the RITS complex. In fission yeast this complex contains argonaute, a chromodomain protein Chp1, and a protein called Tas3 of unknown function. As a consequence, the induction and spread of heterochromatic regions requires the argonaute and RdRP proteins. Indeed, deletion of these genes in the fission yeast *S. pombe* disrupts histone methylation and centromere formation, causing slow or stalled anaphase during cell division. In some cases, similar processes associated with histone modification have been observed to transcriptionally upregulate genes.

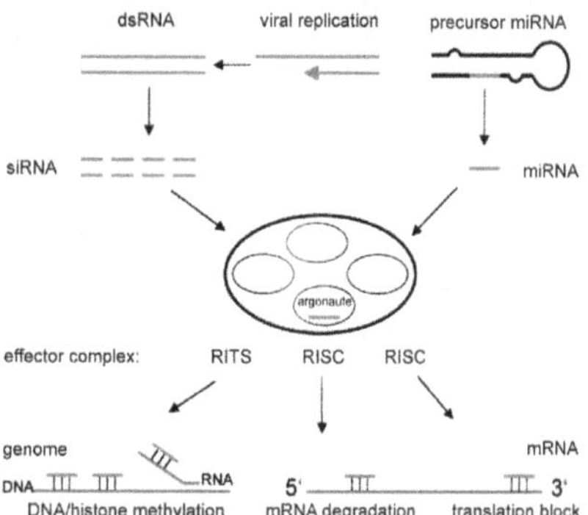

The enzyme dicer trims double stranded RNA, to form small interfering RNA or microRNA. These processed RNAs are incorporated into the RNA-induced silencing complex (RISC), which targets messenger RNA to prevent translation.

The mechanism by which the RITS complex induces heterochromatin formation and organization is not well understood. Most studies have focused on the mating-type region in fission yeast, which may not be representative of activities in other genomic regions/organisms. In maintenance of existing heterochromatin regions, RITS forms a complex with siRNAs complementary to the local genes and stably binds local methylated histones, acting co-transcriptionally to degrade any nascent pre-mRNA transcripts that are initiated by RNA polymerase. The formation of such a heterochromatin region, though not its maintenance, is dicer-dependent, presumably because dicer is required to generate the initial complement of siRNAs that target subsequent transcripts. Heterochromatin maintenance has been suggested to function as a self-reinforcing feedback loop, as new siRNAs are formed from the occasional nascent transcripts by RdRP for incorporation into local RITS complexes. The relevance of observations from fission yeast mating-type regions and centromeres to mammals is not clear, as heterochromatin maintenance in mammalian cells may be independent of the components of the RNAi pathway.

Crosstalk with RNA Editing

The type of RNA editing that is most prevalent in higher eukaryotes converts adenosine nucleotides into inosine in dsRNAs via the enzyme adenosine deaminase (ADAR). It was originally proposed in 2000 that the RNAi and A→I RNA editing pathways might compete for a common dsRNA substrate. Some pre-miRNAs do undergo A→I RNA editing and this mechanism may regulate the processing and expression of mature miRNAs. Furthermore, at least one mammalian ADAR can sequester siRNAs from RNAi pathway components. Further support for this model comes from studies on ADAR-null *C. elegans* strains indicating that A→I RNA editing may counteract RNAi silencing of endogenous genes and transgenes.

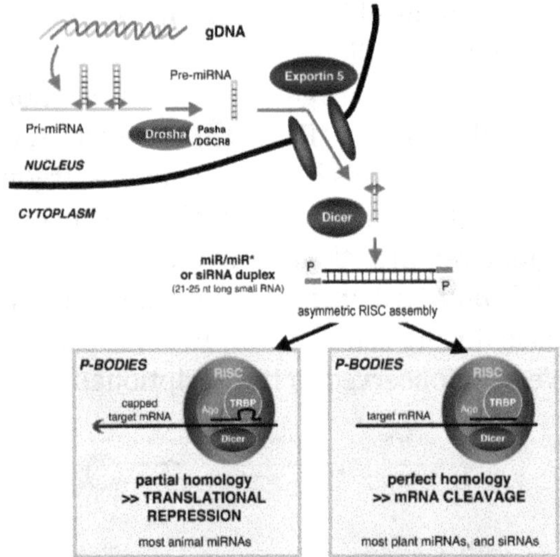

Illustration of the major differences between plant and animal gene silencing. Natively expressed microRNA or exogenous small interfering RNA is processed by dicer and integrated into the RISC complex, which mediates gene silencing.

Variation Among Organisms

Organisms vary in their ability to take up foreign dsRNA and use it in the RNAi pathway. The effects of RNA interference can be both systemic and heritable in plants and *C. elegans*, although not in *Drosophila* or mammals. In plants, RNAi is thought to propagate by the transfer of siRNAs between cells through plasmodesmata (channels in the cell walls that enable communication and transport). Heritability comes from methylation of promoters targeted by RNAi; the new methylation pattern is copied in each new generation of the cell. A broad general distinction between plants and animals lies in the targeting of endogenously produced miRNAs; in plants, miRNAs are usually perfectly or nearly perfectly complementary to their target genes and induce direct mRNA cleavage by RISC, while animals' miRNAs tend to be more divergent in sequence and induce translational repression. This translational effect may be produced by inhibiting the interactions of translation initiation factors with the messenger RNA's polyadenine tail.

Some eukaryotic protozoa such as *Leishmania major* and *Trypanosoma cruzi* lack the RNAi pathway entirely. Most or all of the components are also missing in some fungi, most notably the model organism *Saccharomyces cerevisiae*. The presence of RNAi in other budding yeast species such as *Saccharomyces castellii* and *Candida albicans*, further demonstrates that inducing two RNAi-related proteins from *S. castellii* facilitates RNAi in *S. cerevisiae*. That certain ascomycetes and basidiomycetes are missing RNA interference pathways indicates that proteins required for RNA silencing have been lost independently from many fungal lineages, possibly due to the evolution of a novel pathway with similar function, or to the lack of selective advantage in certain niches.

Related Prokaryotic Systems

Gene expression in prokaryotes is influenced by an RNA-based system similar in some respects to RNAi. Here, RNA-encoding genes control mRNA abundance or translation by producing a complementary RNA that anneals to an mRNA. However these regulatory RNAs are not generally considered to be analogous to miRNAs because the dicer enzyme is not involved. It has been suggested that CRISPR interference systems in prokaryotes are analogous to eukaryotic RNA interference systems, although none of the protein components are orthologous.

Biological Functions

Immunity

RNA interference is a vital part of the immune response to viruses and other foreign genetic material, especially in plants where it may also prevent the self-propagation of transposons. Plants such as *Arabidopsis thaliana* express multiple dicer homologs that are specialized to react differently when the plant is exposed to different viruses. Even before the RNAi pathway was fully understood, it was known that induced gene silencing in plants could spread throughout the plant in a systemic effect and could be transferred from stock to scion plants via grafting. This phenomenon has since been recognized as a feature of the plant adaptive immune system and allows the entire plant to respond to a virus after an initial localized encounter. In response, many plant viruses have evolved elaborate mechanisms to suppress the RNAi response. These include viral proteins that bind short double-stranded RNA fragments with single-stranded overhang ends, such as those produced by dicer. Some plant genomes also express endogenous siRNAs in response to infection by specific types of bacteria. These effects may be part of a generalized response to pathogens that downregulates any metabolic process in the host that aids the infection process.

Although animals generally express fewer variants of the dicer enzyme than plants, RNAi in some animals produces an antiviral response. In both juvenile and adult *Drosophila*, RNA interference is important in antiviral innate immunity and is active against pathogens such as Drosophila X virus. A similar role in immunity may operate in *C. elegans*, as argonaute proteins are upregulated in response to viruses and worms that overexpress components of the RNAi pathway are resistant to viral infection.

The role of RNA interference in mammalian innate immunity is poorly understood, and relatively little data is available. However, the existence of viruses that encode genes able to suppress the RNAi response in mammalian cells may be evidence in favour of an RNAi-dependent mammalian immune response, although this hypothesis has been challenged as poorly substantiated. Maillard et al. and Li et al. provide evidence for the existence of a functional antiviral RNAi pathway in mammalian cells. Other functions for RNAi in mammalian viruses also exist, such as miRNAs expressed by the herpes virus that may act as heterochromatin organization triggers to mediate viral latency.

Downregulation of Genes

Endogenously expressed miRNAs, including both intronic and intergenic miRNAs, are most important in translational repression and in the regulation of development, especially on the timing of morphogenesis and the maintenance of undifferentiated or incompletely differentiated cell types such as stem cells. The role of endogenously expressed miRNA in downregulating gene expression was first described in *C. elegans* in 1993. In plants this function was discovered when the "JAW microRNA" of *Arabidopsis* was shown to be involved in the regulation of several genes that control plant shape. In plants, the majority of genes regulated by miRNAs are transcription factors; thus miRNA activity is particularly wide-ranging and regulates entire gene networks during development by modulating the expression of key regulatory genes, including transcription factors as well as F-box proteins. In many organisms, including humans, miRNAs are linked to the formation of tumors and dysregulation of the cell cycle. Here, miRNAs can function as both oncogenes and tumor suppressors.

Upregulation of Genes

RNA sequences (siRNA and miRNA) that are complementary to parts of a promoter can increase gene transcription, a phenomenon dubbed RNA activation. Part of the mechanism for how these RNA upregulate genes is known: dicer and argonaute are involved, possibly via histone demethylation. miRNAs have been proposed to upregulate their target genes upon cell cycle arrest, via unknown mechanisms.

Evolution

Based on parsimony-based phylogenetic analysis, the most recent common ancestor of all eukaryotes most likely already possessed an early RNA interference pathway; the absence of the pathway in certain eukaryotes is thought to be a derived characteristic. This ancestral RNAi system probably contained at least one dicer-like protein, one argonaute, one PIWI protein, and an RNA-dependent RNA polymerase that may also have played other cellular roles. A large-scale comparative genomics study likewise indicates that the eukaryotic crown group already possessed these components, which may then have had closer functional associations with generalized RNA degradation systems such as the exosome. This study also suggests that the RNA-binding argonaute protein family, which is shared among eukaryotes, most archaea, and at least some bacteria (such as *Aquifex aeolicus*), is homologous to and originally evolved from components of the translation initiation system.

The ancestral function of the RNAi system is generally agreed to have been immune defense against exogenous genetic elements such as transposons and viral genomes. Related functions such as histone modification may have already been present in the ancestor of modern eukaryotes, although other functions such as regulation of development by miRNA are thought to have evolved later.

RNA interference genes, as components of the antiviral innate immune system in many eukaryotes, are involved in an evolutionary arms race with viral genes. Some viruses have evolved mechanisms for suppressing the RNAi response in their host cells, particularly for plant viruses. Studies of evolutionary rates in *Drosophila* have shown that genes in the RNAi pathway are subject to strong directional selection and are among the fastest-evolving genes in the *Drosophila* genome.

Applications

Gene Knockdown

The RNA interference pathway is often exploited in experimental biology to study the function of genes in cell culture and *in vivo* in model organisms. Double-stranded RNA is synthesized with a sequence complementary to a gene of interest and introduced into a cell or organism, where it is recognized as exogenous genetic material and activates the RNAi pathway. Using this mechanism, researchers can cause a drastic decrease in the expression of a targeted gene. Studying the effects of this decrease can show the physiological role of the gene product. Since RNAi may not totally abolish expression of the gene, this technique is sometimes referred as a "knockdown", to distinguish it from "knockout" procedures in which expression of a gene is entirely eliminated.

Extensive efforts in computational biology have been directed toward the design of successful dsRNA reagents that maximize gene knockdown but minimize "off-target" effects. Off-target effects arise when an introduced RNA has a base sequence that can pair with and thus reduce the expression of multiple genes. Such problems occur more frequently when the dsRNA contains repetitive sequences. It has been estimated from studying the genomes of humans, *C. elegans* and *S. pombe* that about 10% of possible siRNAs have substantial off-target effects. A multitude of software tools have been developed implementing algorithms for the design of general mammal-specific, and virus-specific siRNAs that are automatically checked for possible cross-reactivity.

Depending on the organism and experimental system, the exogenous RNA may be a long strand designed to be cleaved by dicer, or short RNAs designed to serve as siRNA substrates. In most mammalian cells, shorter RNAs are used because long double-stranded RNA molecules induce the mammalian interferon response, a form of innate immunity that reacts nonspecifically to foreign genetic material. Mouse oocytes and cells from early mouse embryos lack this reaction to exogenous dsRNA and are therefore a common model system for studying mammalian gene-knockdown effects. Specialized laboratory techniques have also been developed to improve the utility of RNAi in mammalian systems by avoiding the direct introduction of siRNA, for example, by stable transfection with a plasmid encoding the appropriate sequence from which siRNAs can be transcribed, or by more elaborate lentiviral vector systems allowing the inducible activation or deactivation of transcription, known as *conditional RNAi*.

Functional Genomics

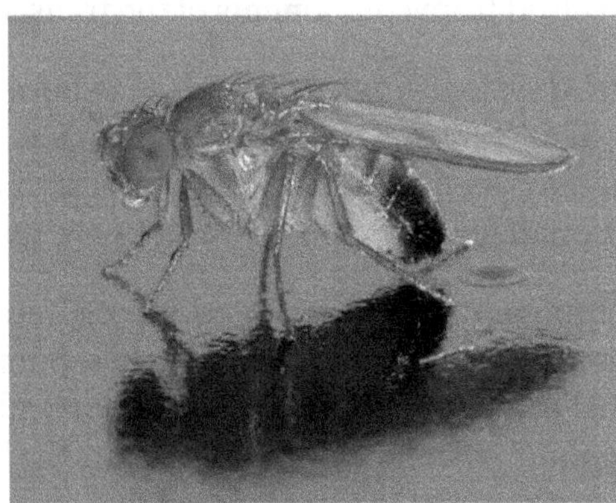

A normal adult Drosophila fly, a common model organism used in RNAi experiments.

Most functional genomics applications of RNAi in animals have used *C. elegans* and *Drosophila*, as these are the common model organisms in which RNAi is most effective. *C. elegans* is particularly useful for RNAi research for two reasons: firstly, the effects of gene silencing are generally heritable, and secondly because delivery of the dsRNA is extremely simple. Through a mechanism whose details are poorly understood, bacteria such as *E. coli* that carry the desired dsRNA can be fed to the worms and will transfer their RNA payload to the worm via the intestinal tract. This "delivery by feeding" is just as effective at inducing gene silencing as more costly and time-consuming delivery methods, such as soaking the worms in dsRNA solution and injecting dsRNA into the gonads. Although delivery is more difficult in most other organisms, efforts are also underway to undertake large-scale genomic screening applications in cell culture with mammalian cells.

Approaches to the design of genome-wide RNAi libraries can require more sophistication than the design of a single siRNA for a defined set of experimental conditions. Artificial neural networks are frequently used to design siRNA libraries and to predict their likely efficiency at gene knockdown. Mass genomic screening is widely seen as a promising method for genome annotation and has triggered the development of high-throughput screening methods based on microarrays. However, the utility of these screens and the ability of techniques developed on model organisms to generalize to even closely related species has been questioned, for example from *C. elegans* to related parasitic nematodes.

Functional genomics using RNAi is a particularly attractive technique for genomic mapping and annotation in plants because many plants are polyploid, which presents substantial challenges for more traditional genetic engineering methods. For example, RNAi has been successfully used for functional genomics studies in bread wheat (which is hexaploid) as well as more common plant model systems *Arabidopsis* and maize.

Medicine

It may be possible to exploit RNA interference in therapy. Although it is difficult to introduce long dsRNA strands into mammalian cells due to the interferon response, the use of short interfering RNA has been more successful. Among the first applications to reach clinical trials were in the treatment of macular degeneration and respiratory syncytial virus. RNAi has also been shown to be effective in reversing induced liver failure in mouse models.

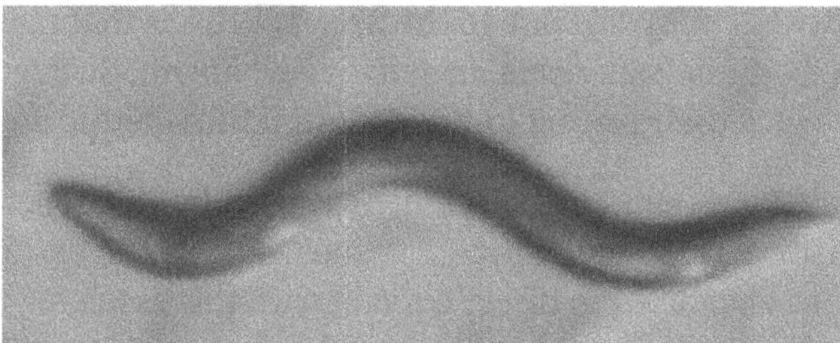

An adult C. elegans worm, grown under RNAi suppression of a nuclear hormone receptor involved in desaturase regulation. These worms have abnormal fatty acid metabolism but are viable and fertile.

Antiviral

Potential antiviral therapies include topical microbicide treatments that use RNAi to treat infection (at Harvard Medical School; in mice, so far) by herpes simplex virus type 2 and the inhibition of viral gene expression in cancerous cells, knockdown of host receptors and coreceptors for HIV, the silencing of hepatitis A and hepatitis B genes, silencing of influenza gene expression, and inhibition of measles viral replication. Potential treatments for neurodegenerative diseases have also been proposed, with particular attention to polyglutamine diseases such as Huntington's disease.

RNA interference-based applications are being developed to target persistent HIV-1 infection. Viruses like HIV-1 are particularly difficult targets for RNAi-attack because they are escape-prone, which requires combinatorial RNAi strategies to prevent viral escape.

Cancer

RNA interference is also a promising way to treat cancers by silencing genes differentially upregulated in tumor cells or genes involved in cell division. A key area of research in the use of RNAi for clinical applications is the development of a safe delivery method, which to date has involved mainly viral vector systems similar to those suggested for gene therapy.

Due to safety concerns with viral vectors, nonviral delivery methods, typically employing lipid-based or polymeric vectors, are also promising candidates. Computational modeling of nonviral siRNA delivery paired with *in vitro* and *in vivo* gene knockdown studies elucidated the temporal behavior of RNAi in these systems. The model used an input bolus dose of siRNA and computationally and experimentally showed that knockdown duration was dependent mainly on the doubling time of the cells to which siRNA was delivered, while peak knockdown depended primarily on the delivered dose. Kinetic considerations of RNAi are imperative to safe and effective dosing schedules as nonviral methods of inducing RNAi continue to be developed.

Safety

Despite the proliferation of promising cell culture studies for RNAi-based drugs, some concern has been raised regarding the safety of RNA interference, especially the potential for "off-target" effects in which a gene with a coincidentally similar sequence to the targeted gene is also repressed. A computational genomics study estimated that the error rate of off-target interactions is about 10%. One major study of liver disease in mice reported that 23 out of 49 distinct RNAi treatment protocols resulted in death. Researchers hypothesized this alarmingly high rate to be the result of "oversaturation" of the dsRNA pathway, due to the use of shRNAs that have to be processed in the nucleus and exported to the cytoplasm using an active mechanism. Such considerations are under active investigation, to reduce their impact in the potential therapeutic applications.

RNAi in vivo delivery to tissues still eludes science—especially to tissues deep within the body. RNAi delivery is only easily accessible to surface tissues such as the eye and respiratory tract. In these instances, siRNA has been used in direct contact with the tissue for transport. The resulting RNAi successfully focused on target genes. When delivering siRNA to deep tissues, the siRNA must be protected from nucleases, but targeting specific areas becomes the main difficulty. This difficulty has been combatted with high dosage levels of siRNA to ensure the tissues have been reached, however in these cases hepatotoxicity was reported.

Biotechnology

RNA interference has been used for applications in biotechnology and is nearing commercialization in others. RNAi has developed many novel crops such as nicotinefree tobacco, decaffeinated coffee, nutrient fortified and hypoallergenic crops. The genetically engineered Arctic apples are near close to receive US approval. The apples were produced by RNAi suppression of PPO (polyphenol oxidase) gene making apple varieties that will not undergo browning after being sliced. PPO-silenced apples are unable to convert chlorogenic acid into quinone product.

There are several opportunities for the applications of RNAi in crop science for its improvement such as stress tolerance and enhanced nutritional level. RNAi will prove its potential for inhibition of photorespiration to enhance the productivity of C3 plants. This knockdown technology may be useful in inducing early flowering, delayed ripening, delayed senescence, breaking dormancy, stress-free plants, overcoming self-sterility, etc.

Foods

RNAi has been used to genetically engineer plants to produce lower levels of natural plant toxins. Such techniques take advantage of the stable and heritable RNAi phenotype in plant stocks. Cotton seeds are rich in dietary protein but naturally contain the toxic terpenoid product gossypol, making them unsuitable for human consumption. RNAi has been used to produce cotton stocks whose seeds contain reduced levels of delta-cadinene synthase, a key enzyme in gossypol production, without affecting the enzyme's production in other parts of the plant, where gossypol is itself important in preventing damage from plant pests. Similar efforts have been directed toward the reduction of the cyanogenic natural product linamarin in cassava plants.

No plant products that use RNAi-based genetic engineering have yet exited the experimental stage. Development efforts have successfully reduced the levels of allergens in tomato plants and fortification of plants such as tomatoes with dietary antioxidants. Previous commercial products, including the Flavr Savr tomato and two cultivars of ringspot-resistant papaya, were originally developed using antisense technology but likely exploited the RNAi pathway.

Other Crops

Another effort decreased the precursors of likely carcinogens in tobacco plants. Other plant traits that have been engineered in the laboratory include the production of non-narcotic natural products by the opium poppy and resistance to common plant viruses.

Insecticide

RNAi is under development as an insecticide, employing multiple approaches, including genetic engineering and topical application. Cells in the midgut of many larvae take up the molecules and help spread the signal throughout the insect's body.

RNAi has varying effects in different species of Lepidoptera (butterflies and moths). Possibly because their saliva is better at breaking down RNA, the cotton bollworm, the beet armyworm and the Asiatic rice borer have so far not been proven susceptible to RNAi by feeding.

To develop resistance to RNAi, the western corn rootworm would have to change the genetic sequence of its Snf7 gene at multiple sites. Combining multiple strategies, such as engineering the protein Cry, derived from a bacterium called Bacillus thuringiensis (Bt), and RNAi in one plant delay the onset of resistance.

Transgenic Plants

Transgenic crops have been made to express small bits of RNA, carefully chosen to silence crucial genes in target pests. RNAs exist that affect only insects that have specific genetic sequences. In 2009 a study showed RNAs that could kill any one of four fruit fly species while not harming the other three.

In 2012 Syngenta bought Belgian RNAi firm Devgen for $522 million and Monsanto paid $29.2 million for the exclusive rights to intellectual property from Alnylam Pharmaceuticals. The International Potato Center in Lima, Peru is looking for genes to target in the sweet potato weevil, a beetle whose larvae ravage sweet potatoes globally. Other researchers are trying to silence genes in ants, caterpillars and pollen beetles. Monsanto will likely be first to market, with a transgenic corn seed that expresses dsRNA based on gene Snf7 from the western corn rootworm, a beetle whose larvae annually cause one billion dollars in damage in the United States alone. A 2012 paper showed that silencing Snf7 stunts larval growth, killing them within days. In 2013 the same team showed that the RNA affects very few other species.

Topical

Alternatively dsRNA can be supplied without genetic engineering. One approach is to add them to irrigation water. The molecules are absorbed into the plants' vascular system and poison insects feeding on them. Another approach involves spraying RNA like a conventional pesticide. This would allow faster adaptation to resistance. Such approaches would require low cost sources of RNAs that do not currently exist.

Genome-Scale Screening

Genome-scale RNAi research relies on high-throughput screening (HTS) technology. RNAi HTS technology allows genome-wide loss-of-function screening and is broadly used in the identification of genes associated with specific phenotypes. This technology has been hailed as the second genomics wave, following the first genomics wave of gene expression microarray and single nucleotide polymorphism discovery platforms. One major advantage of genome-scale RNAi screening is its ability to simultaneously interrogate thousands of genes. With the ability to generate a large amount of data per experiment, genome-scale RNAi screening has led to an explosion data generation rates. Exploiting such large data sets is a fundamental challenge, requiring suitable statistics/ bioinformatics methods. The basic process of cell-based RNAi screening includes the choice of an RNAi library, robust and stable cell types, transfection with RNAi agents, treatment/incubation, signal detection, analysis and identification of important genes or therapeutical targets.

History

Example petunia plants in which genes for pigmentation are silenced by RNAi. The left plant is wild-type; the right plants contain transgenes that induce suppression of both transgene and endogenous gene expression, giving rise to the unpigmented white areas of the flower.

The discovery of RNAi was preceded first by observations of transcriptional inhibition by antisense RNA expressed in transgenic plants, and more directly by reports of unexpected outcomes in experiments performed by plant scientists in the United States and the Netherlands in the early 1990s. In an attempt to alter flower colors in petunias, researchers introduced additional copies of a gene encoding chalcone synthase, a key enzyme for flower pigmentation into petunia plants of normally pink or violet flower color. The overexpressed gene was expected to result in darker flowers, but instead produced less pigmented, fully or partially white flowers, indicating that the activity of chalcone synthase had been substantially decreased; in fact, both the endogenous genes and the transgenes were downregulated in the white flowers. Soon after, a related event termed *quelling* was noted in the fungus *Neurospora crassa*, although it was not immediately recognized as related. Further investigation of the phenomenon in plants indicated that the downregulation was due to post-transcriptional inhibition of gene expression via an increased rate of mRNA degradation. This phenomenon was called *co-suppression of gene expression*, but the molecular mechanism remained unknown.

Not long after, plant virologists working on improving plant resistance to viral diseases observed a similar unexpected phenomenon. While it was known that plants expressing virus-specific proteins showed enhanced tolerance or resistance to viral infection, it was not expected that plants carrying only short, non-coding regions of viral RNA sequences would show similar levels of protection. Researchers believed that viral RNA produced by transgenes could also inhibit viral replication. The reverse experiment, in which short sequences of plant genes were introduced into viruses, showed that the targeted gene was suppressed in an infected plant. This phenomenon was labeled "virus-induced gene silencing" (VIGS), and the set of such phenomena were collectively called *post transcriptional gene silencing*.

After these initial observations in plants, laboratories searched for this phenomenon in other organisms. Craig C. Mello and Andrew Fire's 1998 *Nature* paper reported a potent gene silencing effect after injecting double stranded RNA into *C. elegans*. In investigating the regulation of muscle protein production, they observed that neither mRNA nor antisense RNA injections had an effect on protein production, but double-stranded RNA successfully silenced the targeted gene. As a result of this work, they coined the term *RNAi*. This discovery represented the first identification of the causative agent for the phenomenon. Fire and Mello were awarded the 2006 Nobel Prize in Physiology or Medicine.

References

- Lodish H, Berk A, Matsudaira P, Kaiser CA, Krieger M, Scott MP, Zipurksy SL, Darnell J (2004). Molecular Cell Biology (5th ed.). WH Freeman: New York, NY. ISBN 978-0-7167-4366-8.
- Berkhout, B; ter Brake, O (2010). "RNAi Gene Therapy to Control HIV-1 Infection". RNA Interference and Viruses: Current Innovations and Future Trends. Caister Academic Press. ISBN 978-1-904455-56-1.
- Matson RS (2005). Applying genomic and proteomic microarray technology in drug discovery. CRC Press. ISBN 0-8493-1469-0.
- Zhang XHD (2011). Optimal High-Throughput Screening: Practical Experimental Design and Data Analysis for Genome-scale RNAi Research. Cambridge University Press. ISBN 978-0-521-73444-8.
- Mol JN, van der Krol AR (1991). Antisense nucleic acids and proteins: fundamentals and applications. M. Dekker. pp. 4, 136. ISBN 0-8247-8516-9.

7

Sub-Disciplines of Proteomics

Proteomics has a number of branches; some of the sub-disciplines are phosphoproteomics, degradomics and neuroprotemoics. Phosphoproteomics is the branch of proteomics that helps in the description and in the characterization of proteins. The chapter strategically encompasses and incorporates the major sub-disciplines of proteomics, providing a complete understanding.

Phosphoproteomics

Phosphoproteomics is a branch of proteomics that identifies, catalogs, and characterizes proteins containing a phosphate group as a post-translational modification. Phosphorylation is a key reversible modification that regulates protein function, subcellular localization, complex formation, degradation of proteins and therefore cell signalling networks. With all of these modification results, it is assumed that up to 30% of all proteins may be phosphorylated, some multiple times.

Compared to expression analysis, phosphoproteomics provides two additional layers of information. First, it provides clues on what protein or pathway might be activated because a change in phosphorylation status almost always reflects a change in protein activity. Second, it indicates what proteins might be potential drug targets as exemplified by the kinase inhibitor Gleevec. While phosphoproteomics will greatly expand knowledge about the numbers and types of phosphoproteins, its greatest promise is the rapid analysis of entire phosphorylation based signalling networks.

Overview of Phosphoproteomic Analysis

A sample large-scale phosphoproteomic analysis includes

1. Cultured cells undergo SILAC encoding.
2. Cells are stimulated with factor of interest (e.g. growth factor, hormone).
3. Stimulation can occur for various lengths of time for temporal analysis.
4. Cells are lysed and enzymatically digested.
5. Peptides are separated using ion exchange chromatography.
6. Phosphopeptides are enriched using phosphospecific antibodies, immobilized metal affinity chromatography or titanium dioxide (TiO_2) chromatography.
7. Phosphopeptides are analyzed using mass spectrometry.
8. Peptides are sequenced and analyzed.

Tools and Methods

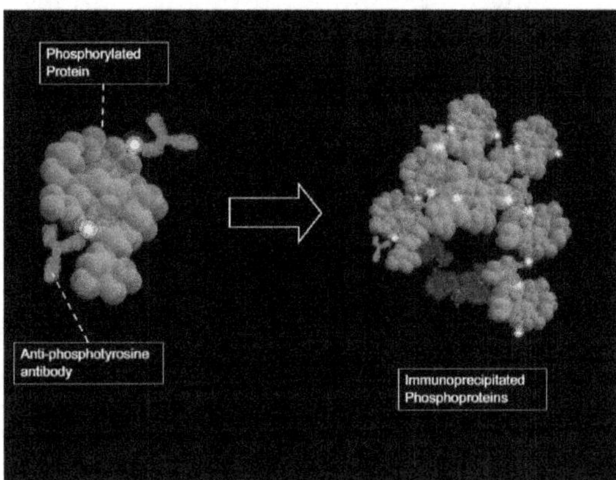

Method of phosphoprotein purification by immunoprecipitation with anti-phosohotyrosine antibodies

The analysis of the entire complement of phosphorylated proteins in a cell is certainly a feasible option. This is due to the optimization of enrichment protocols for phosphoproteins and phosphopeptides, better fractionation techniques using chromatography, and improvement of methods to selectively visualize phosphorylated residues using mass spectrometry. Although the current procedures for phosphoproteomic analysis are greatly improved, there is still sample loss and inconsistencies with regards to sample preparation, enrichment, and instrumentation. Bioinformatics tools and biological sequence databases are also necessary for high-throughput phosphoproteomic studies.

Enrichment Strategies

Previous procedures to isolate phosphorylated proteins included radioactive labeling with P-labeled ATP followed by SDS polyacrylamide gel electrophoresis or thin layer chromatography. These traditional methods are inefficient because it is impossible to obtain large amounts of proteins required for phosphorylation analysis. Therefore, the current and simplest methods to enrich phosphoproteins are affinity purification using phosphospecific antibodies, immobilized metal affinity chromatography (IMAC), strong cation exchange (SCX) chromatography, or titanium dioxide chromatography. Antiphosphotyrosine antibodies have been proven very successful in purification, but fewer reports have been published using antibodies against phosphoserine- or phosphothreonine-containing proteins. IMAC enrichment is based on phosphate affinity for immobilized metal chelated to the resin. SCX separates phosphorylated from non-phosphorylated peptides based on the negatively charged phosphate group. Titanium dioxide chromatography is a newer technique that requires significantly less column preparation time. Many phosphoproteomic studies use a combination of these enrichment strategies to obtain the purest sample possible.

Mass Spectrometry Analysis

Mass spectrometry is currently the best method to adequately compare pairs of protein samples. The two main procedures to perform this task are using isotope-coded affinity tags (ICAT) and stable isotopic amino acids in cell culture (SILAC). In the ICAT procedure samples are labeled individually after isolation with mass-coded reagents that modify cysteine residues. In SILAC, cells are cultured separately in the presence of different isotopically labeled amino acids for several cell divisions allowing cellular proteins to incorporate the label. Mass spectrometry is subsequently used to identify phosphoserine, phosphothreonine, and phosphotyrosine-containing peptides.

Phosphoproteomics in the Study of Signal Transduction

Intracellular signal transduction is primarily mediated by the reversible phosphorylation of various signalling molecules by enzymes dubbed kinases. Kinases transfer phosphate groups from ATP to specific serine, threonine or tyrosine residues of target molecules. The resultant phosphorylated protein may have altered activity level, subcellular localization or tertiary structure.

Phosphoproteomic analyses are ideal for the study of the dynamics of signalling networks. In one study design, cells are exposed to SILAC labelling and then stimulated by a specific growth factor. The cells are collected at various timepoints, and the lysates are combined for analysis by tandem MS. This allows experimenters to track the phosphorylation state of many phosphoproteins in the cell over time. The ability to measure the global phosphorylation state of many proteins at various time points makes this approach much more powerful than traditional biochemical methods for analyzing signalling network behavior.

One study was able to simultaneously measure the fold-change in phosphorylation state of 127 proteins between unstimulated and EphrinB1-stimulated cells. Of these 127 proteins, 40 showed increased phosphorylation with stimulation by EphrinB1. The researchers were able to use this information in combination with previously published data to construct a signal transduction network for the proteins downstream of the EphB2 receptor.

Another recent phosphoproteomic study included large-scale identification and quantification of phosphorylation events triggered by the anti-diuretic hormone vasopressin in kidney collecting duct. A total of 714 phosphorylation sites on 223 unique phosphoproteins were identified, including three novel phosphorylation sites in the vasopressin-sensitive water channel aquaporin-2 (AQP2).

Phosphoproteomics in the Study of Cancer

Since the inception of phosphoproteomics, cancer research has focused on changes to the phosphoproteome during tumor development. Phosphoproteins could be cancer markers useful to cancer diagnostics and therapeutics. In fact, research has shown that there are distinct phosphotyrosine proteomes of breast and liver tumors. There is also evidence of hyperphosphorylation at tyrosine residues in breast tumors but not in normal tissues. Findings like these suggest that it is possible to mine the tumor phosphoproteome for potential biomarkers.

Increasing amounts of data are available suggesting that distinctive phosphoproteins exist in various tumors and that phosphorylation profiling could be used to fingerprint cancers from different origins. In addition, systematic cataloguing of tumor-specific phosphoproteins in individual patients could reveal multiple causative players during cancer formation. By correlating this experimental data to clinical data such as drug response and disease outcome, potential cancer markers could be identified for diagnosis, prognosis, prediction of drug response, and potential drug targets.

Limitations

While phosphoproteomics has greatly expanded knowledge about the numbers and types of phosphoproteins, along with their role in signaling networks, there are still a few limitations to these techniques. To begin with, isolation methods such as anti-phosphotyrosine antibodies do not distinguish between isolating tyrosine-phosphorylated proteins and proteins associated with tyrosine-phosphorylated proteins. Therefore, even though phosphorylation dependent protein-protein interactions are very important, it is important to remember that a protein detected by this method is not necessarily a direct substrate of any tyrosine kinase. Only by digesting the samples before immunoprecipitation can isolation of only phosphoproteins and temporal profiles of individual phosphorylation sites be produced. Another limitation is that some relevant proteins will likely be missed since no extraction condition is all encompassing. It is possible that proteins with low stoichiometry of phosphorylation, in very low abundance, or phosphorylated as a target for rapid degradation will be lost.

Degradomics

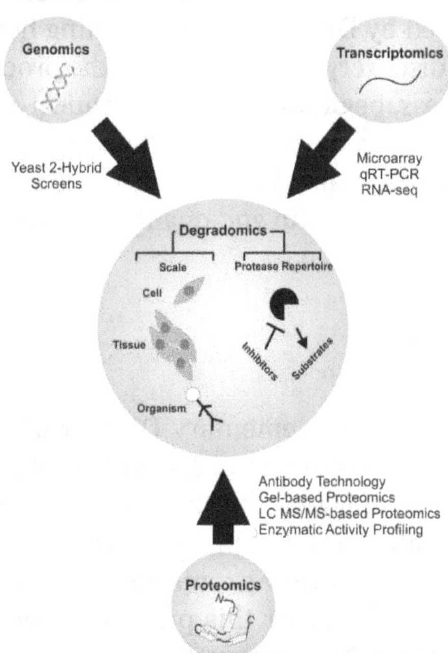

Representation of the relationship of degradomics to genomic, transcriptomic, and proteomic research approaches.

Degradomics is a sub-discipline of biology encompassing all the genomic and proteomic approaches devoted to the study of proteases, their inhibitors, and their substrates on a system-wide scale. This includes the analysis of the protease and protease-substrate repertoires, also called "protease degradomes.". The scope of these degradomes can range from cell, tissue, and organism-wide scales.

Background

As the second largest class of enzymes behind ubiquitin ligases and responsible for ~2% of any organism's genes, proteases have drawn the attention of biologists to develop a field aimed to identify and quantify their roles in biology. First coined in 2000 by the Overall Lab in McQuibban et al, degradomics was described as linking proteases to substrates on a proteome basis. The discoveries of novel roles for proteases and breakthroughs in protease-substrate discovery would be summarized later by Dr. Carlos Lopez-Otin and Dr. Chris Overall, introducing degradomics on a system-wide scale. They collated the current and emerging techniques available to describe proteolysis. By drawing attention to how proteolysis serves as an additional irreversible mechanism by which cells could achieve control over biological processes, they outlined the necessity of studying proteases for their functional relevance in processing bioactive molecules. These bioactive molecules play roles in coagulation, complement activation, DNA replication, cell-cycle control, cellular proliferation and migration, hemostasis, immunity, and apoptosis. The degradome was broken down into two concepts, the first referring the entire profile of proteases expressed under by a cell, tissue, or organism under defined circumstances. The second definition applies specifically to the full substrate repertoire of a certain protease in a cell, tissue, or organism.

Dr. Overall's group would go on to annotate the complete human and mouse protease-inhibitor degradomes in 2003. As the complexity of proteolytic networks was uncovered, more thorough descriptions of the degradome and ways to study it became necessary. In 2007, Dr. Overall updated the field with a review co-authored by Dr. Carl Blobel detailing how advanced methods were revolutionizing protease-substrate discovery. They described the process of linking proteases to their substrates on a step by step process, beginning with biochemical and proteomic discovery, validation using cellular based assays, and progressing to whole organism levels using animal models. More recently, as technology and techniques have advanced, the Overall Lab and others have continued to direct the field using more powerful and quantitative techniques.

Transcriptomic Methods

Gene Microarrays

Traditionally DNA microarrays use complementary DNA or oligonucleotide probes to analyze messenger RNA (mRNA) from genes of interest. Extracted total RNA serves as a template for complementary DNA (cDNA) that is tagged with fluorescent probes before being allowed to hybridize to the microarray for visualization. For proteases, specific probes for protease genes and their inhibitors have been developed to view expression patterns on the mRNA transcript level. The two platforms currently available for this purpose come from corporate and academic sources. Affymetrix's Hu/Mu ProtIn Microarray uses 516 and 456 probe sets to evaluate human and murine proteases, inhibitors, and interactors respectively. CLIP-CHIP™, developed by the Overall Lab, is a complete protease and inhibitor DNA microarray for all 1561 human and murine proteases, non-proteolytic homologues, and their inhibitors. Both of these tools allow comparison of expression patterns between normal and diseased samples and tissues. Unfortunately, as transcript levels often fail to reflect protein expression levels, gene microarrays are limited in representing protein in samples. In addition, proteases recruited from remote sources like nearby tissues are ignored by these DNA based arrays, reiterating the need for protein based methods to confirm the presence and activity of functional enzymes when transcriptome analysis is performed.

Quantitative Real-time Polymerase Chain Reaction (qRT-PCR)

A more sensitive approach to transcript analysis of a gene is quantitative real-time polymerase chain reaction (qRT-PCR), which has also seen action in quantifying protease mRNA levels. Again, total RNA is extracted and used to generate cDNA for PCR amplification. As long as there is a specific primer for a protease, protease inhibitor, or interacting molecule, qRT-PCR can serve as a highly sensitive method to detect miniscule amounts of mRNA copy numbers per cell. A drawback that separates it from microarray analysis is its limited scope: a microarray can handle parallel analysis of multiple genes while qRT-PCR must amplify one mRNA for analysis at a time. It also suffers from the same limitation of microarray analysis regarding the lack of correlation between transcript and protein levels. However, its sensitivity lends it as a useful tool in validating microarray findings and quantifying specific protease transcripts of interest.

RNA Sequencing (RNA-seq)

Whole transcriptome shotgun sequencing (WTSS) is the latest in gene expression studies, using next generation sequencing (NGS) to quantify RNA in samples on a high throughput scale. As biology trends toward using RNA-seq over microarray analysis in evaluating the transcriptome, so does degradomics. The field adapts the approach to analyzing the presence and quantity of transcripts of proteases, their substrates, and their inhibitors. While developed microarrays remain a major workhorse in studying gene expression in degradomics, its limitations of cross hybridization and dynamic range issues suggest RNA-seq will take a larger role as costs decrease and analysis improves.

Genomic Methods

Yeast Two-hybrid Screens

Yeast two-hybrid analyses have been adapted for protease-substrate discovery. As protease exosites play roles in protein-protein recognition and interaction, biologists have used exosites as tools to screen for protease interactors and potential substrates. These protease exosite scanning assays use protease exosites as bait to scan a cDNA library for possible interacting partners.

Another early adaptation of yeast two-hybrid screening in protease-substrate discovery is Inactive-catalytic-domain capture (ICDC). This approach attempts to avoid the limitation of protease exosite scanning, which fails to account for any substrates that do not require exosites to for recognition before cleavage. The bait for these assays are immobilized catalytically inactive mutant protease domains that cannot cleave and release their substrates once bound.

While useful in early degradomic studies, the limitations of adapted yeast two-hybrid screens have forced the field to move on to higher-throughput approaches for protease-substrate discovery. Their high rate of false positives and negatives, inability to recognize complex interactions, lack of biologically compartmentalization, and failure to account for post-translational modifications necessary for protein-protein interactions hamper their usefulness. Thus they have been largely replaced by proteomic methods as technology has improved.

Proteomic Methods

Protease-specific Arrays

A protease specific protein array based on immobilized antibodies designed to capture specific proteases from biological samples offers a step up in analysis of protein levels beyond transcript expression. Capture antibodies spotted to nitrocellulose membranes can bind proteases in complex mixtures which have been pre-incubated and bound by detection antibodies allowing for parallel analysis of relative protease levels. These arrays offer parallelization of protein levels over traditional western blot. Unfortunately, these assays fail to provide insight on enzymatic function for proteases and suffer similar drawbacks to western blots regarding reliable quantification.

Immunohistochemistry Approaches

As an antibody technique, immunohistochemistry (IHC) allows for validating protein presence. It, and immunocytochemistry, allow for surveying the localization of proteases on a tissue or cellular scale respectively. It also can evaluate for the localization of cleavage products using monoclonal antibodies raised against neo-epitopes of cleavage sites produced by protease processing. Unfortunately, in addition to providing little functional information, IHC is also non-quantitative, making it an unappealing option for describing degradomics on system-wide scales.

Gel-based Proteomic Methods

Two-dimensional Polyacrylamide Electrophoresis (2D-PAGE) gels historically compared intensities from protease treated and untreated sample spots in order to identify possible candidate substrates. A more recent improvement of this technique, fluorescent 2D difference gel electrophoresis (2D-DIGE), attempts to control standardization between gels for relative quantification. Differentially labelling protease-treated and untreated samples with either Cy3 or Cy5, pooling said samples, and analyzing them together by 2D-PAGE allows substrate and cleavage products to be studied from the fluorescent gel. The spots corresponding to potentially substrate and cleavage products can be later elucidated using Mass Spectrometry or Edman Sequencing. The biggest drawbacks to using these techniques relate to the chemistry of the technique itself and its lack of sensitivity. As they rely on PAGE gels, extremely large, small, highly hydrophobic, acidic, or basic molecules will not be visualized.

Mass Spectrometry-based Proteomic Methods

Conventional shotgun proteomics identification of low abundance proteins in samples remains limited despite advances in Mass Spectrometry (MS) technology. While abundant proteins can be easily detected, possible protease substrates of biological significance, such as cytokines, can be easily overlooked due to their low abundance. Most pre-clearing strategies designed to correct this also risk losing low abundant proteins, thus techniques designed specifically to target protease substrates for identification have been developed. These techniques have coalesced into a new field of positional proteomics or terminomics aimed at identifying protein N- or C-terminal modifications of protease substrates. Terminomic approaches including Terminal Amine Isotopic Labeling of Substrates (TAILS) N-Terminomics, Combined FRActional Diagonal Chromatography (COFRADIC), and C-Terminomics add the level of stringency to conventional shotgun proteomics necessary to make them workhorse of degradomics.

TAILS, or "N-Terminomics," was designed and developed by the Overall Lab to overcome the functional limitations of conventional proteomics by enriching both mature N-terminal peptides and newly generated N-terminal peptides of proteins produced by protease activity. Formaldehyde or isobaric tags including Isotope-coded Affinity Tags (ICAT), 4 to 8 plex Isobaric tag for relative and absolute quantification (iTRAQ), or 10plex Tandem mass tags (TMT) block primary amines prior to trypsin digestion of proteome samples. The main step of the process is the negative selection of newly generated trypsin peptides using a specialized polymer. The polymer ignores the unreactive primary amines blocked by their tags, allowing them to be separated from trypsin generated peptides by ultrafiltration for Liquid Chromatography Tandem Mass Spectrometry (LC-MS/MS) analysis. These mature and neo-N-termini will differ in ratios between the protease treated versus untreated samples and make up the proteolytic fingerprint of a protease. TAILS is also compatible with Stable isotope labeling by amino acids in cell culture (SILAC).

COFRADIC was the earliest technique to capitalize on negative selection to enrich for protein N-termini. Sample proteins are first blocked by reduction and alkylation at their primary amines before endopeptidase treatment. Its negative selection method relies on strong cation exchange chromatography (SCX) to enrich for peptides representing N- and C-termini of proteins based on differences in peptide charge and pH. Additional orthogonal chromatography treatments change the biochemical character of the peptides for further enrichment before final LC-MS/MS analysis. Groups have continued to adapt and improve this technology for protease-substrate discovery.

C-terminomics has always been complicated due to the chemical nature of its targets. Carboxyl groups are less reactive than primary amines, making C-terminomic techniques more complex than established N-terminomic approaches. However, adapted TAILS and COFRADIC workflows have been developed specifically to study the C-termini of proteins. Recently, the Overall Lab tackled another difficulty of C-terminomics, using endopeptidase LysargiNase™ to generate C-termini carrying N-terminal lysine or arginine residues. Previously, C-termini lacked basic residues after endopeptidase digestion and could be missed in LC-MS/MS workflows.

Another approach designed at further elucidating protease activity is Proteomic Identification of protease Cleavage Sites (PICS). Beginning with a peptide library generated from endopeptidase digestion of a proteome, this technique allows for screening and characterizing the prime- and non-prime specificity for proteases. After digestion, primary amines and sulfhydryl are chemically blocked before digesting the sample again with the desired protease. Now, protease generated primary amines that constitute the prime site of cleavage can be biotinylated and isolated due to their reactivity and analyzed by LC-MS/MS. Non-prime sides sequences left behind must be determined using bioinformatics analysis of the extracted N-termini and full length protein sequences. These prime and non-prime sites give a full picture of protease cleavage site specificity.

Activity-based Profiling

To achieve functional degradomics, the enzymatic activity of proteases must be analyzed. Methods have been developed to distinguish the proteolytic activity of different enzymes in biological samples and separate active proteases from their inactive forms, namely zymogen precursors and those proteases bound by inhibitors. Two techniques are activity-based probes (ABPs) and Proteolytic Signature Peptides (PSPs).

ABP molecules serve as probes to irreversibly bind only to active proteases and ignore their zymogen precursors and inhibited proteases. Placing a reactive group and a recognizable tag feature on the same molecule using a linker moiety gives an ABP molecule its structure. The reactive molecule, designed after protease inhibitor mechanisms, lends ABPs their specificity towards targeting active proteases. Once bound, the reactive group acts much like an irreversible inhibitor to the protease. Depending on the nature of the tag moiety, the ABP-protease complex can then be visualized or retrieved from biological samples for further studies of localization and quantification. Limitations including difficult production, specificity, stability, and toxicity hamper ABP development but these probes have proved useful in revealing protease biological activity and remain a promising avenue in degradomic technology.

PSPs do not depend on targeting active proteases with tagged compounds but rather on quantitative proteomics using stable isotope labeled standard peptides. Standard peptides synthesized from amino acids labeled with stable isotope atoms serve as internal standards for serial dilutions of a sample. These allow for later absolute quantification of a proteins and post-translational modifications by mass spectrometry. This technique has been adapted to absolute quantification of proteases, deciphering both activity states and total amounts in biological samples. This is thanks to trypsin treatment for mass spectrometry generating peptides specific to inactive zymogen precursors, active proteases, or common to both forms. PSPs is one form of standard peptide for absolute quantification and Standard of the Expressed Protease (STEP) is the other. The major difference between the two is PSP sequences are designed to mimic tryptic peptides that contain sequences spanning the zymogen's pro-domain and the protease's final form, whereas STEP sequences match tryptic peptide sequences found in both forms of the protease. This capitalizes on protease activation, where the zymogen's pro-domain is cleaved off and the final form lacks that domain. Experiments using iTRAQ labeling and LC-MS/MS with STEP and PSP peptide internal standards have successfully quantified total and active protease levels in biological samples. One major drawback for this approach is the inability to account for inhibitor bound enzymes. It is also difficult to ensure standard peptides can be generated for this method for each and every protease for study.

Experiments using iTRAQ labeling and LC-MS/MS with STEP and PSP peptide internal standards have successfully quantified total and active protease levels in biological samples. One major drawback for this approach is the inability to account for inhibitor bound enzymes. It is also difficult to ensure standard peptides can be generated for this method for each and every protease for study. However, PSPs hold large potential for translating degradomics into clinical applications, as once a PSP is established it could aid in quantifying proteolytic signature biomarkers in Single Reaction Monitoring (SRM) and Multiple Reaction Monitoring (MRM) type clinical assays.

Bioinformatics

Owing to the increasing complexity of regulation of cellular processes and the roles proteases play in them, bioinformatics continues to be an invaluable tool for degradomics. Software, databases, and projects developed for this purpose have accompanied the advancement in technology. Software developed in the Overall Lab (CLIPPER) statistically evaluates cleavage site candidates determined by degradomic approaches. One web-based data site, WebPICS, incorporates and integrates cleavage site analysis from PICS experiments into MEROPS, the protease database. Another database, Termini oriented protein Function Inferred Database (TopFIND), serves as a knowledge base to integrate protein termini formed by protease processing with functional interpretations. By combining research literature and other biological databases including UniProt, MEROPS, Ensembl, and TisDB, the database comprehensively renders protein termini modifications accessible to a broad scientific community. Using TopFIND, terminal modifications can be identified and visualized across proteins thanks to all available in silico, in vitro, and in vivo findings. Using TopFINDer and Path FINDer software, research findings can be mathematically modelled into a network of pathways regulated by proteases, further contributing to the "protease web".

Significance and Impact

Protease Biology

Thanks to rapid advances in proteomics, genomics, and bioinformatics, protease research has been revolutionized. Degradomics emerged with the concept that proteolysis represents a specific mechanism for achieving cellular control over vital processes beyond control afforded by gene expression and translation and continues to produce the research necessary to understand the complex regulation of biology. Where it was thought extracellular proteases degraded extracellular matrix (ECM), these proteases are now known to target and process a vast array of substrates with diverse roles, redefining protease functions and leading to a shift in interest towards new roles previously unknown to biology. Degradomic studies of human tissue have also contributed to the Human Proteome Project (HPP) of the Human Proteome Organization (HUPO).

Precision Medicine

The protease modulatory web represents opportunities to identify novel biomarkers for disease and targets for drug design. Proteolytic processed N-termini have been proposed as potential biomarkers as disease specific proteolysis has been well studied in pathologies such as inflammation and cancer. Contributions to degradomics have identified numerous characterized and novel protease substrates and continue to lead to speculation of previously unknown protease targets. More recently, proteolytic signatures of cell death have been found using N-terminomic techniques on chemotherapy patient plasma samples. Advancements in SRM and MRM clinical assays also allow for analyzing proteolytic signature biomarkers in patient samples and can be complemented by PSP quantification. Deciphering these networks will aid drug design in understanding which substrates perform useful roles versus harmful ones to determine which should be targeted by drugs.

Neuroproteomics

Neuroproteomics is the study of the protein complexes and species that make up the nervous system. These proteins interact to make the neurons connect in such a way to create the intricacies that nervous system is known for. Neuroproteomics is a complex field that has a long way to go in terms of profiling the entire neuronal proteome. It is a relatively recent field that has many applications in therapy and science. So far, only small subsets of the neuronal proteome have been mapped, and then only when applied to the proteins involved in the synapse.

History

Origins

The word *proteomics* was first used in 1994 by Marc Wilkins as the study of "the protein equivalent of a genome". It is defined as all of the proteins expressed in a biological system under specific physiologic conditions at a certain point in time. It can change with any biochemical alteration, and so it can only be defined under certain conditions. Neuroproteomics is a subset of this field dealing with the complexities and multi-system origin of neurological disease. Neurological function is based on the interactions of many proteins of different origin, and so requires a systematic study of subsystems within its proteomic structure.

Modern Times

Neuroproteomics has the difficult task of defining on a molecular level the pathways of consciousness, senses, and self. Neurological disorders are unique in that they do not always exhibit outward symptoms. Defining the disorders becomes difficult and so neuroproteomics is a step in the right direction of identifying bio-markers that can be used to detect diseases. Not only does the field have to map out the different proteins possible from the genome, but there are many modifications that happen after transcription that affect function as well. Because neurons are such dynamic structures, changing with every action potential that travels through them, neuroproteomics offers the most potential for mapping out the molecular template of their function. Genomics offers a static roadmap of the cell, while proteomics can offer a glimpse into structures smaller than the cell because of its specific nature to each moment in time.

Mechanisms of Use

Protein Separation

In order for neuroproteomics to function correctly, proteins must be separated in terms of the proteome from which they came. For example, one set might be under normal conditions, while another might be under diseased conditions. Proteins are commonly separated using two-dimensional polyacrylamide gel electrophoresis (2D PAGE). For this technique, proteins are run across an immobile gel with a pH gradient until they stop at the point where their net charge is neutral. After separating by charge in one direction, sodium dodecyl sulfate is run in the other direction to separate the proteins by size. A two-dimensional map is created using this technique that can be used to match additional proteins later. One can usually match the function of a protein by identifying in an 2D PAGE in simple proteomics because many intracellular somatic pathways are known. In neuroproteomics, however, many proteins combine to give an end result that may be neurological disease or breakdown. It is necessary then to study each protein individually and find a correlation between the different proteins to determine the cause of a neurological disease. New techniques are being developed that can identify proteins once they are separated out using 2D PAGE.

Protein Identification

Protein separate techniques, such as 2D PAGE, are limitd in that they cannot handle very high or low molecular weight protein species. Alternative methods have been developed to deal with such cases. These include liquid chromatography mass spectrometry along with sodium dodecyl sulfate polyacrylamide gel electrophoresis, or liquid chromatography mass spectrometry run in multiple dimensions. Compared to simple 2D page, liquid chromatography mass spectrometry can handle a larger range of protein species size, but it is limited in the amount of protein sample it handle at once. Liquid chromatography mass spectrometry is also limited in its lack of a reference map from which to work with. Complex algorithms are usually used to analyze the fringe results that occur after a procedure is run. The unknown portions of the protein species are usually not analyzed in favor of familiar proteomes, however. This fact reveals a fault with current technology; new techniques are needed to increase both the specificity and scope of proteome mapping.

Applications

Drug Addiction

It is commonly known that drug addiction involves permanent synaptic plasticity of various neuronal circuits. Neuroproteomics is being applied to study the effect of drug addiction across the synapse. Research is being conducted by isolating distinct regions of the brain in which synaptic transmission takes place and defining the proteome for that particular region. Different stages of drug abuse must be studied, however, in order to map out the progression of protein changes along the course of the drug addiction. These stages include enticement, ingesting, withdrawal, addiction, and removal. It begins with the change in the genome through transcription that occurs due to the abuse of drugs. It continues to identify the most likely proteins to be affected by the drugs and focusing in on that area. For drug addiction, the synapse is the most likely target as it involves communication between neurons. Lack of sensory communication in neurons is often an outward sign of drug abuse, and so neuroproteomics is being applied to find out what proteins are being affected to prevent the transport of neurotransmitters. In particular, the vesicle releasing process is being studied to identify the proteins involved in the synapse during drug abuse. Proteins such as synaptotagmin and synaptobrevin interact to fuse the vesicle into the membrane. Phosphorylation also has its own set of proteins involved that work together to allow the synapse to function properly. Drugs such as morphine change properties such as cell adhesion, neurotransmitter volume, and synaptic traffic. After significant morphine application, tyrosine kinases received less phosphorylation and thus send fewer signals inside the cell. These receptor proteins are unable to initiate the intracellular signaling processes that enable the neuron to live, and necrosis or apoptosis may be the result. With more and more neurons affected along this chain of cell death, permanent loss of sensory or motor function may be the result. By identifying the proteins that are changed with drug abuse, neuroproteomics may give clinicians even earlier biomarkers to test for to prevent permanent neurological damage.

Recently, a novel terminology (Psychoproteomics) has been coined by the University of Florida researchers from Dr. Mark S Gold Lab. Kobeissy et al. defined Psychoproteomics as integral proteomics approach dedicated to studying proteomic changes in the field of psychiatric disorders, particularly substance-and drug-abuse neurotoxicity.

Brain Injury

Traumatic brain injury is defined as a "direct physical impact or trauma to the head followed by a dynamic series of injury and repair events". Recently, neuroproteomics have been applied to studying the disability that over 5.4 million Americans live with. In addition to physically injuring the brain tissue, traumatic brain injury induces the release of glutamate that interacts with ionotropic glutamate receptors (iGluRs). These glutamate receptors acidify the surrounding intracranial fluid, causing further injury on the molecular level to nearby neurons. The death of the surrounding neurons is induced through normal apoptosis mechanisms, and it is this cycle that is being studied with neuroproteomics. Three different cysteine protease derivatives are involved in the apoptotic pathway induced by the acidic environment triggered by glutamate. These cysteine proteases include calpain, caspase, and cathepsin. These three proteins are examples of detectable signs of traumatic brain injury that are much more specific than temperature, oxygen level, or intracranial pressure. Proteomics thus also offers a tracking mechanism by which researchers can monitor the progression of traumatic brain injury, or a chronic disease such as Alzheimer's or Parkinson's. Especially in Parkinson's, in which neurotransmitters play a large role, recent proteomic research has involved the study of synaptotagmin. Synaptotagmin is involved in the calcium-induced budding of vesicle containing neurotransmitters from the presynaptic membrane. By studying the intracellular mechanisms involved in neural apoptosis after traumatic brain injury, researchers can create a map that genetic changes can follow later on.

Nerve Growth

One group of researchers applied the field of neuroproteomics to examine how different proteins affect the initial growth of neuritis. The experiment compared the protein activity of control neurons with the activity of neurons treated with nerve growth factor (NGF) and JNJ460, an "immunophilin ligand." JNJ460 is an offspring of another drug that is used to prevent immune attack when organs are transplanted. It is not an immunosuppressant, however, but rather it acts as a shield against microglia. NGF promotes neuron viability and differentiation by binding to TrkA, a tyrosine receptor kinase. This receptor is important in initiating intracellular metabolic pathways, including Ras, Rak, and MAP kinase.

Protein differentiation was measured in each cell sample with and without treatment by NGF and JNJ460. A peptide mixture was made by washing off unbound portions of the amino acid sequence in a reverse column. The resulting mixture was then suspended a peptide mixture in a bath of cation exchange fluid. The proteins were identified by splicing them with trypsin and then searching through the results of passing the product through a mass spectrometer. This applies a form of liquid chromatography mass spectrometry to identify proteins in the mixture

JNJ460 treatment resulted in an increase in "signal transduction" proteins, while NGF resulted in an increase in proteins associated with the ribosome and synthesis of other proteins. JNJ460 also resulted in more structural proteins associated with intercellular growth, such as actin, myosin, and troponin. With NGF treatment, cells increased protein synthesis and creation of ribosomes. This method allows the analysis of all of the protein patterns overall, rather than a single change in an amino acid. Western blots confirmed the results, according to the researchers, though the changes in proteins were not as obvious in their protocol.

The main significance to these findings are that JNJ460 are NGF are distinct processes that both control the protein output of the cell. JNJ460 resulted in increased neuronal size and stability while NGF resulted in increased membrane proteins. When combined together, they significantly increase a neuron's chance of growth. While JNJ460 may "prime" some parts of the cell for NGF treatment, they do not work together. JNJ460 is thought to interact with Schwann cells in regenerating actin and myosin, which are key players in axonal growth. NGF helps the neuron grow as a whole. These two proteins do not play a part in communication with other neurons, however. They merely increase the size of the membrane down which a signal can be sent. Other neurotrophic factor proteomes are needed to guide neurons to each other to create synapses.

Limitations

The broad scope of the available raw neuronal proteins to map requires that initial studies be focused on small areas of the neurons. When taking samples, there are a few places that interest neurologists most. The most important place to start for neurologists is the plasma membrane. This is where most of the communication between neurons takes place. The proteins being mapped here include ion channels, neurotransmitter receptors, and molecule transporters. Along the plasma membrane, the proteins involved in creating cholesterol-rich lipid rafts are being studied because they have been shown to be crucial for glutamate uptake during the initial stages of neuron formation. As mentioned before, vesicle proteins are also being studied closely because they are involved in disease. Collecting samples to study, however, requires special consideration to ensure that the reproducibility of the samples is not compromised. When taking a global sample of one area of the brain for example, proteins that are ubiquitous and relatively unimportant show up very clear in the SDS PAGE. Other unexplored, more specific proteins barely show up and are therefore ignored. It is usually necessary to divide up the plasma membrane proteome, for example, into subproteomes characterized by specific function. This allows these more specific classes of peptides to show up more clearly. In a way, dividing into subproteomes is simply applying a magnifying lens to a specific section of a global proteome's SDS PAGE map. This method seems to be most effective when applied to each cellular organelle separately. Mitochondrial proteins, for example, which are more effective at transporting electrons across its membrane, can be specifically targeted effectively in order to match their electron-transporting ability to their amino acid sequence.

References

- Vizovisek M., Vidmar R., Fonovic M., Turk B. Current trends and challenges in proteomic identification of protease substrates. Biochimie. 122, 77-87 (2016).
- Marino G., Eckhard U., Overall C. M. Protein Termini and Their Modifications Revealed by Positional Proteomics. ACS Chem. Biol. 10, 1754-1764 (2015).
- Huesgen P.F., Lange P.F., Overall C.M. Ensembles of protein termini and specific proteolytic signatures as candidate biomarkers of disease. Proteom. Clin. Appl., 8 338–350 (2014).
- Chen S.-H., Chen C.-R., Chen S.-H., Li D.-T., Hsu J.-L. Improved N(α)-acetylated peptide enrichment following dimethyl labeling and SCX J. Proteom. Res. 12, 3277–3287 (2013).
- auf dem Keller U., Overall C.M. CLIPPER: an add-on to the Trans-Proteomic Pipeline for the automated analysis of TAILS N-terminomics data. Biol Chem. 393(12):1477-83 (2012).
- Huesgen P.F., Overall C.M. N- and C-terminal degradomics: new approaches to reveal biological roles for plant proteases from substrate identification Physiol. Plant, 145, 5–17 (2012).
- Lange P.F., Huesgen P.F., Overall C.M. TopFIND 2.0 — Linking protein termini with proteolytic processing and modifications altering protein function. Nucleic Acids Res. 40 ,D351 – 361 (2012).

Permissions

All chapters in this book are published with permission under the Creative Commons Attribution Share Alike License or equivalent. Every chapter published in this book has been scrutinized by our experts. Their significance has been extensively debated. The topics covered herein carry significant information for a comprehensive understanding. They may even be implemented as practical applications or may be referred to as a beginning point for further studies.

We would like to thank the editorial team for lending their expertise to make the book truly unique. They have played a crucial role in the development of this book. Without their invaluable contributions this book wouldn't have been possible. They have made vital efforts to compile up to date information on the varied aspects of this subject to make this book a valuable addition to the collection of many professionals and students.

This book was conceptualized with the vision of imparting up-to-date and integrated information in this field. To ensure the same, a matchless editorial board was set up. Every individual on the board went through rigorous rounds of assessment to prove their worth. After which they invested a large part of their time researching and compiling the most relevant data for our readers.

The editorial board has been involved in producing this book since its inception. They have spent rigorous hours researching and exploring the diverse topics which have resulted in the successful publishing of this book. They have passed on their knowledge of decades through this book. To expedite this challenging task, the publisher supported the team at every step. A small team of assistant editors was also appointed to further simplify the editing procedure and attain best results for the readers.

Apart from the editorial board, the designing team has also invested a significant amount of their time in understanding the subject and creating the most relevant covers. They scrutinized every image to scout for the most suitable representation of the subject and create an appropriate cover for the book.

The publishing team has been an ardent support to the editorial, designing and production team. Their endless efforts to recruit the best for this project, has resulted in the accomplishment of this book. They are a veteran in the field of academics and their pool of knowledge is as vast as their experience in printing. Their expertise and guidance has proved useful at every step. Their uncompromising quality standards have made this book an exceptional effort. Their encouragement from time to time has been an inspiration for everyone.

The publisher and the editorial board hope that this book will prove to be a valuable piece of knowledge for students, practitioners and scholars across the globe.

Index

www.ingramcontent.com/pod-product-compliance
Lightning Source LLC
Jackson TN
JSHW052208130125
77033JS00004B/219

* 9 7 8 1 6 3 5 4 9 2 3 9 2 *